Elementary Number Theory

A series of books in the mathematical sciences
Editor: Victor Klee

Elementary Number Theory

Second Edition

Underwood Dudley

DePauw University

W. H. FREEMAN AND COMPANY
San Francisco

Library of Congress Cataloging in Publication Data

Dudley, Underwood.
 Elementary number theory.

 (A Series of books in the mathematical sciences)
 Bibliography: p.
 Includes index.
 1. Numbers, Theory of. I. Title.
QA241.D79 1978 512'.7 78-5661
ISBN 0-7167-0076-X

AMS Classification: Primary 1001
 Number Theory: Elementary exposition (collegiate level)

Elementary Number Theory, Second Edition

Printed in the United States of America

1 2 3 4 5 6 7 8 9

Contents

Preface

Mathematics exists mainly to give us power and control over the physical world, but it has always been so fascinating that it was studied for its own sake. Number theory is that sort of mathematics: it is of no use in building bridges, and civilization would carry on much as usual if all of its theorems were to disappear, nevertheless it has been studied and valued since the time of Pythagoras. That greatest of mathematicians Carl Friedrich Gauss called it "The Queen of Mathematics," and "Everybody's Mathematics" is what the contemporary mathematician Ivan Niven calls it. The reason for its appeal is that the subject matter—numbers—is part of everyone's experience, and the things that can be found out about them are interesting, curious, or surprising, and the ways they are found can be delightful: clean lines of logic, with sustained tension and satisfying resolutions.

A course in number theory can do several things for a student. It can acquaint him or her with ideas no student of mathematics should be ignorant of. More important, it is an example of the mathematical style of thinking—problem, deduction, solution—in a system where the problems are not unnatural or artificial. Most important, it can help to diminish the feeling that many students have, consciously or not, that mathematics is a collection of formulas and that to solve a problem you need only find the appropriate formula.

This text has been designed for a one-semester or one-quarter course in number theory, with minimal prerequisites. The reader is not required to know any mathematics except elementary algebra and the properties of the real numbers. Nevertheless, the average student does not find number theory easy because it involves understanding new ideas and the proofs of theorems. I have tried to make the proofs detailed enough to be clear, and I have included numerical examples, not only to illustrate the ideas, but to show the fascination of playing with numbers, which is how many of the ideas originated.

I have included an introduction to most of the topics of elementary number theory. In Sections 1 through 5 the fundamental properties of the integers and congruences are developed, and in Section 6 proofs of Fermat's and Wilson's theorems are given. The number theoretic functions d, σ, and ϕ are introduced in Sections 7 to 9. Sections 10 to 12 culminate in the quadratic reciprocity theorem. There follow three more or less independent blocks of material: the representation of numbers (Sections 13 to 15), diophantine equations (16 to 20), and primes (21 and 22). Because I think that problems are especially important and interesting in number theory, Section 23 consists of 260 additional problems, some classified by section and some arranged without regard to topic.

There are three appendixes. Appendix A, Proof by Induction, should be read when and if necessary. Because computers integrate naturally with number theory, Appendix B presents problems for which a computer can be programmed. Appendix C contains a table that makes it easy to factor any positive integer less than 10,000.

Because I believe that the best way to learn mathematics is to try to solve problems, the text includes almost a thousand exercises and problems. I attribute the success of the first edition not to the exposition—after all, the proofs were already known—but to the problems, and the problem lists have been revised, deleting unsuccessful problems and including new ones that may be more successful. The exercises interrupt the text and can be used in several ways: the student may do them as he reads the material for the first time; he may return to them later to check on his understanding of material already studied; or the instructor may include them in his exposition. Some of the exercises and problems are computational and some classical, but many are more or less original, and a few, I think, are startling. Number theory problems can be difficult because inspiration is sometimes necessary to find a solution, and inspiration cannot be had to

order. A student should not expect to be able to conquer all of the problems and should not feel discouraged if some are baffling. There is benefit in trying to solve problems whether a solution is found or not. I. A. Barnett has written [1] "To discover mathematical talent, there is no better course in elementary mathematics than number theory. Any student who can work the exercises in a modern text in number theory should be encouraged to pursue a mathematical career."

Answers are provided where appropriate for exercises and odd-numbered problems—those marked with an asterisk. Comments are given for those problems marked with a dagger. Although there are more problems than a student could solve in one semester, they should be treated as part of the text, to be read even if not solved. Sometimes they may be more interesting than the material on which they are based.

The first edition contained many errors, and I want to thank the many people who pointed them out and suggested improvements. These errors have all been removed, but inevitably new ones have been added. I hope that when the reader finds one, he will feel pleased with his acuteness rather than annoyed with the author. Corrections will be welcomed.

Underwood Dudley
May 1978

Elementary Number Theory

Section

Integers

The subject matter of number theory is numbers, and a large part of number theory is devoted to studying the properties of the integers— that is, the numbers . . . , $-2, -1, 0, 1, 2, \ldots$. Usually the integers are used merely to convey information (3 apples, \$32, $17x^2 + 9$), with no consideration of their properties. When counting apples, dollars, or x^2's, it is immaterial how many divisors 3 has, whether 32 is prime or not, or that 17 can be written as the sum of the squares of two integers. But the integers are so basic a part of mathematics that they have been thought worthy of study for their own sake. The same situation arises elsewhere: the number theorist is comparable to the linguist, who studies words and their properties, independent of their meaning.

There are many replies to the question, "Why study numbers?" Here are some that have been given:

Because teacher says you must.

Because you won't graduate if you don't.

Because you have to take *something*.

Because it gives your mind valuable training in thinking logically.

Because numbers might be interesting.

Because numbers are a fundamental part of man's mental universe and hence worth looking into.

Because some of the most powerful human minds that ever existed were concerned with numbers, and what powerful minds study is worth studying.

Because you want to know all about numbers: what makes them work, and what they do.

Because mathematics contains some beautiful things, and someone told you that number theory contained some of the most beautiful— and few of the most ugly—things.

Because it is fun.

Let us begin.

In this section, and until further notice, lower case *italic* letters will invariably denote integers. We will take as known and use freely the usual properties of addition, subtraction, multiplication, division, and order for the integers. We also use in this section an important property of the integers—a property that you may not be consciously aware of because it is not stated explicitly as the others. It is the *least-integer principle:* a nonempty set of integers that is bounded below contains a smallest element. There is the corresponding *greatest-integer principle:* a nonempty set of integers that is bounded above contains a largest element.

We will say that a **divides** b (written $a|b$) if and only if there is an integer d such that $ad = b$. For example, $2|6$, $12|60$, $17|17$, $-5|50$, and $8|-24$. If a does not divide b, we will write $a \nmid b$. For example, $4 \nmid 2$ and $3 \nmid 4$.

* **Exercise 1.†** Which integers divide zero?

Exercise 2. Show that if $a|b$ and $b|c$, then $a|c$.

As a sample of the sort of properties that division has, we prove

Lemma 1. If $d|a$ and $d|b$, then $d|(a + b)$.

Proof. From the definition, we know that there are integers q and r such that

$$dq = a \quad \text{and} \quad dr = b.$$

† Answers to selected exercises, those preceded by an asterisk (*), are provided on pp. 226ff.

Thus

$$a + b = d(q + r),$$

so from the definition again, $d \mid (a + b)$.

In the same way, we can prove

Lemma 2. If $d \mid a_1, d \mid a_2, \ldots, d \mid a_n$, then $d \mid (c_1 a_1 + c_2 a_2 + \cdots + c_n a_n)$ for any integers c_1, c_2, \ldots, c_n.

Proof. From the definition, there are integers q_1, q_2, \ldots, q_n such that $a_1 = dq_1, a_2 = dq_2, \ldots, a_n = dq_n$. Thus

$$c_1 a_1 + c_2 a_2 + \cdots + c_n a_n = d(c_1 q_1 + c_2 q_2 + \cdots + c_n q_n),$$

and from the definition again, $d \mid c_1 a_1 + c_2 a_2 + \cdots + c_n a_n$.

Exercise 3. Prove that if $d \mid a$ then $d \mid ca$ for any integer c.

As an application of Lemma 2, let us see if it is possible to have 100 coins, made up of c pennies, d dimes, and q quarters, be worth exactly $5.00. If it is possible, then

$$c + d + q = 100$$

and

$$c + 10d + 25q = 500.$$

Subtract the first equation from the second and we get $9d + 24q = 400$. Since $3 \mid 9$ and $3 \mid 24$, Lemma 2 says that $3 \mid 9d + 24q$. That is, $3 \mid 400$. But that is impossible, so having exactly $5.00 is impossible with 100 pennies, dimes, and quarters. There are, however, five different ways of getting $4.99, and later we will develop a method for finding them.

Fractions are not as natural as integers, and there seems to be a human tendency to avoid them. For example, we divide a gallon into quarts, a quart into pints, and a pint into ounces so that we can always measure with integer multiples of some unit. Finding a unit common to different measures was a problem which would arise naturally in commerce—if 15 Athenian drachmas are worth 18 drachmas from Miletus, how many Athenian drachmas are equivalent to 60 Miletian drachmas? That is one reason why the Euclidean Algorithm for finding

greatest common divisors was part of Euclid's *Elements,* written around 300 BC. The rest of this section will be devoted to the greatest common divisor and its properties, which we will use constantly later. We say that d is the **greatest common divisor** of a and b (written $d = (a, b)$) if and only if

(i) $d|a$ and $d|b$, and

(ii) if $c|a$ and $c|b$, then $c \le d$.

Condition (i) says that d is a common divisor of a and b, and (ii) says that it is the greatest such divisor. For example, $(2, 6) = 2$ and $(3, 4) = 1$. Note that if a and b are not both zero, then the set of common divisors of a and b is a set of integers that is bounded above by the largest of a, b, $-a$, and $-b$. Hence, from the greatest-integer principle for the integers, the set has a largest element, so the greatest common divisor of a and b exists and is unique. Note that $(0, 0)$ is not defined, and that if (a, b) is defined, then it is positive. In fact, $(a, b) \ge 1$ because $1|a$ and $1|b$ for all a and b.

* **Exercise 4.** What are $(4, 14)$, $(5, 15)$, and $(6, 16)$?

* **Exercise 5.** What is $(n, 1)$, where n is any positive integer? What is $(n, 0)$?

* **Exercise 6.** If d is a positive integer, what is (d, nd)?

As an exercise in applying the definition of greatest common divisor, we will prove the following theorem, which we will use often later:

Theorem 1. If $(a, b) = d$, then $(a/d, b/d) = 1$.

Proof. Suppose that $c = (a/d, b/d)$. We want to show that $c = 1$. We will do this by showing that $c \le 1$ and $c \ge 1$. The latter inequality follows from the fact that c is the greatest common divisor of two integers, and as we have noted, every greatest common divisor is greater than or equal to 1. To show that $c \le 1$, we use the facts that $c|(a/d)$ and $c|(b/d)$. We then know that there are integers q and r such that $cq = a/d$ and $cr = b/d$. That is,

$$(cd)q = a \qquad \text{and} \qquad (cd)r = b.$$

These equations show that cd is a common divisor of a and b. It is thus no greater than the *greatest* common divisor of a and b, and this is d.

Thus $cd \leq d$. Since d is positive, this gives $c \leq 1$. Hence $c = 1$, as was to be proved.

If $(a, b) = 1$, then we will say that a and b are **relatively prime,** for a reason that will become clear in the section on unique factorization.

When a and b are small, it is often possible to see what (a, b) is by inspection. When a and b are large, this is no longer possible. The Euclidean Algorithm makes it easy, but first we need.

Theorem 2. The Division Algorithm. Given positive integers a and b, $b \neq 0$, there exist unique integers q and r, with $0 \leq r < b$ such that

$$a = bq + r.$$

Proof. Consider the set of integers $\{a, a - b, a - 2b, a - 3b, \ldots\}$. It contains a subset of nonnegative integers which is nonempty (because a is positive) and bounded below (by 0); from the least-integer principle, it contains a smallest element. Let it be $a - qb$. This number is not negative and it is less than b, because if it were greater than b it would not be the smallest nonnegative element in the set: $a - (q + 1)b$ would be.

$$a - (q+1)b \quad 0 \quad a - qb \quad \cdots \quad a - 2b \quad a - b \quad a$$

Let $r = a - bq$: this construction gives us q and r, and it remains to show that they are unique. Suppose that we have found q, r and q_1, r_1 such that

$$a = bq + r = bq_1 + r_1$$

with $0 \leq r < b$ and $0 \leq r_1 < b$. Subtracting, we have

(1) $$0 = b(q - q_1) + (r - r_1).$$

Since b divides the left-hand side of this equation and the first term on the right-hand side, it divides the other term:

$$b \,|\, (r - r_1).$$

But since $0 \leq r < b$ and $0 \leq r_1 < b$, we have

$$-b < r - r_1 < b.$$

The only multiple of b between $-b$ and b is zero. Hence $r - r_1 = 0$, and it follows from (1) that $q - q_1 = 0$ too. Hence the numbers q and r in the theorem are unique.

Although the theorem was stated only for positive integers a and b, because it is most often applied for positive integers, nowhere in the proof did we need a to be positive. Moreover, if b is negative, the theorem is true if $0 \le r < b$ is replaced with $0 \le r < -b$; you are invited to reread the proof and verify that this is so.

* **Exercise 7.** What are q and r if $a = 75$ and $b = 24$? If $a = 75$ and $b = 25$?

Theorem 2, combined with the next lemma, will give the Euclidean Algorithm.

Lemma 3. If $a = bq + r$, then $(a, b) = (b, r)$.

Proof. Let $d = (a, b)$. We know that since $d \mid a$ and $d \mid b$, it follows from $a = bq + r$ that $d \mid r$. Thus d is a common divisor of b and r. Suppose that c is any common divisor of b and r. We know that $c \mid b$ and $c \mid r$, and it follows from $a = bq + r$ that $c \mid a$. Thus c is a common divisor of a and b, and hence $c \le d$. Both parts of the definition of greatest common divisor are satisfied, and we have $d = (b, r)$.

Exercise 8. Verify that the lemma is true when $a = 16, b = 6$, and $q = 2$.

Let us apply Lemma 3 to find the greatest common divisor of 69 and 21. From $69 = 3 \cdot 21 + 7$ we get $(69, 21) = (21, 7)$, and from $21 = 3 \cdot 7$ we get $(21, 7) = 7$. The ancient Greeks would have found the greatest common measure of these two lengths

by laying the shorter against the longer as many times as possible

and then breaking off the remainder and applying it

until, as in this case, a common measure is found. The formal statement of the process just carried out for a special case is

Theorem 3. The Euclidean Algorithm. If a and b are positive integers, $b \neq 0$, and

$$
\begin{aligned}
a &= bq + r, & 0 &\leq r < b, \\
b &= rq_1 + r_1, & 0 &\leq r_1 < r, \\
r &= r_1 q_2 + r_2, & 0 &\leq r_2 < r_1, \\
& \quad \cdot & & \quad \cdot \\
& \quad \cdot & & \quad \cdot \\
& \quad \cdot & & \quad \cdot \\
r_k &= r_{k+1} q_{k+2} + r_{k+2}, & 0 &\leq r_{k+2} < r_{k+1},
\end{aligned}
$$

then for k large enough, say $k = t$, we have

$$r_{t-1} = r_t q_{t+1},$$

and $(a, b) = r_t$.

Proof. The sequence of nonnegative integers

$$b > r > r_1 > r_2 > \cdots$$

must come to an end. Eventually, one of the remainders will be zero. Suppose that it is r_{t+1}. Then $r_{t-1} = r_t q_{t+1}$. From Lemma 3 applied over and over,

$$(a, b) = (b, r) = (r, r_1) = (r_1, r_2) = \cdots = (r_{t-1}, r_t) = r_t.$$

If either a or b is negative, we can use the fact that $(a, b) = (-a, b) = (a, -b) = (-a, -b)$.

* **Exercise 9.** Calculate $(343, 280)$ and $(578, 442)$.

The computation of $(343, 280) = 7$:

(2) $\qquad\qquad 343 = 1 \cdot 280 + 63$

(3) $\qquad\qquad 280 = 4 \cdot 63 + 28$

(4) $\qquad\qquad\ \ 63 = 2 \cdot 28 + 7$

$\qquad\qquad\qquad\ \ 28 = 4 \cdot 7$

can be worked backward. From (4), $7 = 63 - 2 \cdot 28$. Substitute for 28 from (3): $7 = 63 - 2(280 - 4 \cdot 63) = 9 \cdot 63 - 2 \cdot 280$. Substitute for 63 from (2): $7 = 9(343 - 280) - 2 \cdot 280 = 9 \cdot 343 - 11 \cdot 280$. We have found x and y such that $343x + 280y = 7$, namely $x = 9$ and $y = -11$. What was done in this example can be done in general:

Theorem 4. If $(a, b) = d$, then there are integers x and y such that $ax + by = d$.

Proof. Work the Euclidean Algorithm backward. The details are omitted.

We will find a better method for solving $ax + by = (a, b)$ later, so the computational process in Theorem 4 is not important. What is important is the existence of x and y and not their values. To illustrate the usefulness of Theorem 4, here are three corollaries.

Corollary 1. If $d \mid ab$ and $(d, a) = 1$, then $d \mid b$.

Proof. Since d and a are relatively prime, we know from Theorem 4 that there are integers x and y such that

$$dx + ay = 1.$$

Multiplying this by b, we have

$$d(bx) + (ab)y = b.$$

The term $d(bx)$ can of course be divided by d, and so can $(ab)y$, since d divides ab. Thus d divides the left-hand side of the last equation and hence divides the right-hand side too, which is what we wanted to prove. Note that if d and a are not relatively prime in Corollary 1, then the conclusion is false. For example, $6 \mid 8 \cdot 9$, but $6 \nmid 8$ and $6 \nmid 9$.

Corollary 2. Let $(a, b) = d$, and suppose that $c \mid a$ and $c \mid b$. Then $c \mid d$.

Proof. We know that there are integers x and y such that

$$ax + by = d.$$

Since c divides each term on the left-hand side of this equation, c divides the right-hand side too.

This corollary thus says that every common divisor of a pair of integers is a divisor of their greatest common divisor.

Corollary 3. If $a \mid m$, $b \mid m$, and $(a, b) = 1$, then $ab \mid m$.

Proof. There is an integer q such that $m = bq$, and since $a \mid m$ we have $a \mid bq$. But $(a, b) = 1$, so Corollary 1 says that $a \mid q$. Hence there is an integer r such that $q = ar$, and thus $m = bq = bar$. That shows that $ab \mid m$.

Problems*

* 1. Calculate $(314, 159)$ and $(4144, 7696)$.
 2. Calculate $(3141, 1592)$ and $(10001, 100083)$.
* 3. Find x and y such that $314x + 159y = 1$.
 4. Find x and y such that $4144x + 7696y = 592$.
 5. If $N = abc + 1$, prove that $(N, a) = (N, b) = (N, c) = 1$.
 6. Find two different solutions of $299x + 247y = 13$.
 7. Prove that if $a \mid b$ and $b \mid a$, then $a = b$ or $a = -b$.
 8. Prove that if $a \mid b$ and $a > 0$, then $(a, b) = a$.
 9. Prove that $((a, b), b) = (a, b)$.
 10. (a) Prove that $(n, n + 1) = 1$ for all $n > 0$.
 (b) If $n > 0$, what can $(n, n + 2)$ be?
 11. (a) Prove that $(k, n + k) = 1$ if and only if $(k, n) = 1$.
 * (b) Is it true that $(k, n + k) = d$ if and only if $(k, n) = d$?
 12. Prove: If $a \mid b$ and $c \mid d$, then $ac \mid bd$.
 13. Prove: If $d \mid a$ and $d \mid b$, then $d^2 \mid ab$.
 14. Prove: If $c \mid ab$ and $(c, a) = d$, then $c \mid db$.
 15. (a) If $x^2 + ax + b = 0$ has an integer root, show that it divides b.
 (b) If $x^2 + ax + b = 0$ has a rational root, show that it is in fact an integer.

* Answers to selected odd-numbered problems, those preceded by an asterisk (*), are provided on pp. 231ff. Comments on selected odd-numbered problems, those preceded by a dagger (†), are given on pp. 238ff.

2

Unique Factorization

The aim of this section is to introduce the prime numbers, which are one of the main objects of study in number theory, and to prove the unique factorization theorem for positive integers, which is essential in what comes later. In this section, lower-case italic letters invariably denote *positive* integers.

A **prime** is an integer that is greater than 1 and has no positive divisors other than 1 and itself. An integer that is greater than 1 but is not prime is called **composite.** Thus 2, 3, 5, and 7 are prime, and 4, 6, 8, and 9 are composite. There are also large primes:

$$170,141,183,460,469,231,731,687,303,715,884,105,727$$

is one, and it is clear that there are arbitrarily large composite numbers. Note that we call 1 neither prime nor composite. Although it has no positive divisors other than 1 and itself, including it among the primes would make the statement of some theorems inconvenient, in particular the unique factorization theorem. We will call 1 a **unit.** Thus the set of positive integers can be divided into three classes: the primes, the composites, and a unit.

* **Exercise 1.** How many even primes are there? How many whose last digit is 5?

Our aim is to show that each positive integer can be written as a product of primes—and, moreover, in only one way. We will not count products that differ only in the order of their factors as different factorizations. Thus we will consider each of

$$2^2 \cdot 3 \cdot 7, \qquad 2 \cdot 3 \cdot 7 \cdot 2, \qquad 7 \cdot 3 \cdot 2 \cdot 2$$

to be the same factorization of 84. The primes can thus be used to build, by multiplication, the entire system of positive integers. The first two lemmas that follow will show that every positive integer can be written as a product of primes. Later we will prove the uniqueness of the representation.

Lemma 1. Every integer n, $n > 1$, is divisible by a prime.

Proof. Consider the set of divisors of n which are greater than 1 and less than n. It is either empty or nonempty. If it is empty, then n is prime by definition, and thus has a prime divisor, namely itself. If it is nonempty, then the least-integer principle says that it has a smallest element, call it d. If d had a divisor greater than 1 and less than d, then so would n, but this is impossible because d was the smallest such divisor. Thus d is prime, and n has a prime divisor, namely d.

Lemma 1 can also be proved using the second principle of induction (see Appendix A). The lemma is true by inspection for $n = 2$. Suppose it is true for $n \leq k$. Then either $k + 1$ is prime, in which case we are done, or it is divisible by some number k_1 with $k_1 \leq k$. But from the induction assumption, k_1 is divisible by a prime, and this prime also divides $k + 1$. Again, we are done.

With the aid of Lemma 1, we can prove that every positive integer can be written as a product of primes in at least one way.

Lemma 2. Every integer n, $n > 1$, can be written as a product of primes.

Proof. From Lemma 1, we know that there is a prime p_1 such that $p_1 | n$. That is, $n = p_1 n_1$, where $1 \leq n_1 < n$. If $n_1 = 1$, then we are done: $n = p_1$ is an expression of n as a product of primes. If $n_1 > 1$, then from Lemma 1 again, there is a prime that divides n_1. That is, $n_1 = p_2 n_2$, where p_2 is a prime and $1 \leq n_2 < n_1$. If $n_2 = 1$, again we are done:

$n = p_1p_2$ is written as a product of primes. But if $n_2 > 1$, then Lemma 1 once again says that $n_2 = p_3n_3$, with p_3 a prime and $1 \leq n_3 < n_2$. If $n_3 = 1$, we are done. If not we continue. We will sooner or later come to one of the n_i equal to 1, because $n > n_1 > n_2 > \cdots$ and each n_i is positive; such a sequence cannot continue forever. For some k, we will have $n_k = 1$, in which case $n = p_1p_2 \cdots p_k$ is the desired expression of n as a product of primes. Note that the same prime may occur several times in the product.

Exercise 2 (optional). Construct a proof of Lemma 2 using induction.

* **Exercise 3.** Write prime decompositions for 72 and 480.

Before we show that each positive integer has only one prime decomposition, we will prove an old and elegant theorem:

Theorem 1 (Euclid). There are infinitely many primes.

Proof. Suppose not. Then there are only finitely many primes. Denote them by p_1, p_2, \ldots, p_r. Consider the integer

(1) $$n = p_1p_2 \cdots p_r + 1.$$

From Lemma 1, we see that n is divisible by a prime, and since there are only finitely many primes, it must be one of p_1, p_2, \ldots, p_r. Suppose that it is p_k. Then since

$$p_k | n \quad \text{and} \quad p_k | p_1p_2 \cdots p_r,$$

it divides two of the terms in (1). Consequently it divides the other term in (1); thus $p_k | 1$. This is nonsense: no primes divide 1 because all are greater than 1. This contradiction shows that we started with an incorrect assumption. Since there cannot be only finitely many primes, there are infinitely many.

The table on page 13 shows how adding 1 to $p_1p_2 \cdots p_r$ will always give a prime different from p_1, p_2, \ldots, p_r.

Theorem 1 is strong. We can actually identify only finitely many primes—the largest prime currently known is $2^{19937} - 1$, and we by no means know all of the primes smaller than this one. (A list of all the primes smaller than 10,000,000 fills a large book.) The prime $2^{19937} - 1$ is a very large number: it has more than 6000 digits. Although $2^{19937} - 1$ is a large number, there are infinitely many integers larger than it, and only finitely many smaller. Thus, although we can name only finitely

r	p_r	$p_1 p_2 \cdots p_r + 1$	Prime Divisors
1	2	3	3
2	3	7	7
3	5	31	31
4	7	211	211
5	11	2311	2311
6	13	30031	59, 509
7	17	510511	19, 97, 277
8	19	9699691	347, 27953

many primes, we may be sure that no matter how many we discover, there is always one more that we have yet to find. Before the development of computers, the largest prime known was the comparatively puny 39-digit number displayed at the beginning of this section. Hence if you set out to find a prime larger than $2^{19937} - 1$ without the aid of a machine, you will need a great deal of time to spare—several centuries at the least.

We will show how to construct a table of prime numbers before proving the unique factorization theorem.

Lemma 3. If n is composite, then it has a divisor d such that $1 < d \le n^{1/2}$.

Proof. Since n is composite, there are integers d_1 and d_2 such that $d_1 d_2 = n$ and $1 < d_1 < n,\ 1 < d_2 < n$. If d_1 and d_2 are both larger than $n^{1/2}$, then

$$n = d_1 d_2 > n^{1/2} n^{1/2} = n,$$

which is impossible. Thus, one of d_1 and d_2 must be less than or equal to $n^{1/2}$.

Lemma 4. If n is composite, then it has a prime divisor less than or equal to $n^{1/2}$.

Proof. We know from Lemma 3 that n has a divisor—call it d—such that $1 < d \le n^{1/2}$. From Lemma 1, we know that d has a prime divisor p. Since $p \le d \le n^{1/2}$, the lemma is proved.

Lemma 4 provides the basis for the following method for finding primes, the well-known *Sieve of Eratosthenes,* named after the

Alexandrian mathematician who lived in the third century BC who is also remembered for being the first to estimate the circumference of the earth using geometry. That was a golden century for mathematics: besides Eratosthenes, there was Archimedes, who had one of the most powerful mathematical minds ever, and Euclid, who wrote his geometry book so well that it was used as a textbook for the next two thousand years.

The idea behind the sieve is simple: the primes are the numbers left when all the composites are gone. Then, to find primes remove multiples of 2, of 3, of 5, . . . ; if we stop after removing multiples of N, the numbers remaining between 2 and N^2 are precisely the primes less than N^2. To see this, note that any number remaining is prime, because if it were composite it would have, according to Lemma 4, a prime divisor less than or equal to N, and all the multiples of such primes have been removed.

For example, let us list the integers up to 121 with the multiples of 2 already removed:

2	3	5	7	9	11	13	15	17	19	21	23	25
27	29	31	33	35	37	39	41	43	45	47	49	51
53	55	57	59	61	63	65	67	69	71	73	75	77
79	81	83	85	87	89	91	93	95	97	99	101	103
105	107	109	111	113	115	117	119	121				

Now remove all the multiples of 3—every third number after 3:

2	3	5	7	11	13	17	19	23	25	29	31	35
37	41	43	47	49	53	55	59	61	65	67	71	73
77	79	83	85	89	91	95	97	101	103	107	109	113
115	119	121										

Now remove the multiples of 5, which fall in a pattern: every seventh, third, seventh, third, . . . number; a stencil could be made to pick them out, and such stencils have been made in the past:

2	3	5	7	11	13	17	19	23	29	31	37	41
43	47	49	53	59	61	67	71	73	77	79	83	89
91	97	101	103	107	109	113	119	121				

After 49, the pattern is to have a multiple of 7 every seventh, fourth, seventh, fourth, . . . number; removing them and the multiples of 11, we have

2	3	5	7	11	13	17	19	23	29	31	37	41
43	47	53	59	61	67	71	73	79	83	89	97	101
103	107	109	113									

and these are all the primes less than 121. If there were a composite number in the list, from Lemma 4 it would have a prime divisor less than or equal to 11, and we have removed all of the composite numbers with divisors 2, 3, 5, 7, and 11. To find all of the primes less than 10,000, we would only have to cross out the multiples of the 25 primes less than 100.

Today, any sieving that is necessary is done by computer, but in the nineteenth century, before there were computers, an Austrian astronomer named Kulik constructed an enormous sieve of all the integers up to 100,000,000. It took him 20 years, off and on. All that work was so little valued that the library to which he left his manuscript lost the part that included the integers from 12,642,600 to 22,852,800. The simple sieve idea is quite powerful, and refinements of it—the Selberg Sieve, and the new Large Sieve of Linnik—are producing new results.

The following lemma, proved in Euclid's *Elements,* gives the result that makes unique factorization possible. For the rest of this section, and until further notice, the letters p and q will be reserved for primes.

Lemma 5. If $p \mid ab$, then $p \mid a$ or $p \mid b$.

Proof. Since p is prime, its only positive divisors are 1 and p. Thus $(p, a) = p$ or $(p, a) = 1$. In the first case, $p \mid a$, and we are done. In the second case, Corollary 1 of Section 1 tells us that $p \mid b$, and again we are done.

The next lemma illustrates a common technique: extending a result from 2 to any number using mathematical induction.

Lemma 6. If $p \mid a_1 a_2 \ldots a_k$, then $p \mid a_i$ for some i, $i = 1, 2, \ldots, k$.

Proof. Lemma 6 is true by inspection if $k = 1$, and Lemma 5 shows that it is true if $k = 2$. We will proceed by induction. Suppose that Lemma 6 is true for $k = r$. Suppose that $p \mid a_1 a_2 \ldots a_{r+1}$. Then $p \mid (a_1 a_2 \ldots a_r) a_{r+1}$, and Lemma 5 lets us conclude that $p \mid a_1 a_2 \ldots a_r$ or $p \mid a_{r+1}$. In

the first case, the induction assumption tells us that $p \mid a_i$ for some i, $i = 1, 2, \ldots, r$. In the second case, $p \mid a_i$ for $i = r + 1$. In either case, $p \mid a_i$ for some i, $i = 1, 2, \ldots, r + 1$. Thus, if the lemma is true for $k = r$, it is true for $k = r + 1$, and since it is true for $k = 1$ and $k = 2$, it is true for any positive integer k.

Lemma 7. *If q_1, q_2, \ldots, q_n are primes, and $p \mid q_1 q_2 \cdots q_n$, then $p = q_k$ for some k.*

Proof. From Lemma 6 we know that $p \mid q_k$ for some k. Since p and q_k are primes, $p = q_k$. (The only positive divisors of q_k are 1 and q_k, and p is not 1.)

Theorem 2. The Unique Factorization Theorem. Any positive integer can be written as a product of primes in one and only one way.

Proof. Recall that we agreed to consider as identical all factorizations that differ only in the order of the factors.

We know already from Lemma 2 that any integer n, $n > 1$ can be written as a product of primes. Thus to complete the proof of the theorem, we need to show that n cannot have two such representations. That is, if

$$(2) \qquad n = p_1 p_2 \cdots p_m \qquad \text{and} \qquad n = q_1 q_2 \cdots q_r,$$

then we must show that the same primes appear in each product, and the same number of times, though their order may be different. That is, we must show that the integers p_1, p_2, \ldots, p_m are just a rearrangement of the integers q_1, q_2, \ldots, q_r. From (2) we see that since $p_1 \mid n$,

$$p_1 \mid q_1 q_2 \cdots q_r.$$

From Lemma 7, it follows that $p_1 = q_i$ for some i. If we divide

$$p_1 p_2 \cdots p_m = q_1 q_2 \cdots q_r$$

by the common factor, we have

$$(3) \qquad p_2 p_3 \cdots p_m = q_1 q_2 \cdots q_{i-1} q_{i+1} \cdots q_r.$$

Because p_2 divides the left-hand side of this equation, it also divides the right-hand side. Applying Lemma 7 again, it follows that $p_2 = q_j$ for some $j (j = 1, 2, \ldots, i - 1, i + 1, \ldots, n)$. Cancel this factor from both

sides of (3), and continue the process. Eventually we will find that each p is a q. We cannot run out of q's before all the p's are gone, because we would then have a product of primes equal to 1, which is impossible. If we repeat the argument with the p's and q's interchanged, we see that each q is a p. Thus the numbers p_1, p_2, \ldots, p_m are a rearrangement of q_1, q_2, \ldots, q_r, and the two factorizations differ only in the order of the factors.

The uniqueness of the prime decomposition can also be efficiently proved by induction, though the idea is no different. The theorem is true, by inspection, for $n = 2$. Suppose that it is true for $n \le k$. Suppose that $k + 1$ has two representations:

$$k + 1 = p_1 p_2 \cdots p_m = q_1 q_2 \cdots q_r.$$

As in the last proof, $p_1 = q_i$ for some i, so

$$p_2 p_3 \cdots p_m = q_1 q_2 \cdots q_{i-1} q_{i+1} \cdots q_r.$$

But this number is less than or equal to k, and by the induction assumption, its prime decomposition is unique. Hence the integers $q_1, q_2, \ldots,$ $q_{i-1}, q_{i+1}, \ldots, q_r$ are a rearrangement of p_2, p_3, \ldots, p_m, and since $p_1 = q_i$ the proof is complete.

Because of your long experience with the positive integers (can you remember what it was like not to know what $2 + 3$ was?), you may not find the unique factorization theorem very exciting; you may even think that it is obvious and self-evident. The following example is intended to show that it is not as self-evident as you might think: we will construct a number system in which the unique factorization theorem is not true. Consider the integers 1, 5, 9, 13, 17, . . . ; that is, all integers of the form $4n + 1$, $n = 0, 1, \ldots$. We will call an element of this set *prome* if it has no divisors other than 1 and itself *in the set*. For example, 21 is prome, whereas $25 = 5 \cdot 5$ is not.

* **Exercise 4.** Which members of the set less than 100 are not prome?

In the same way that we proved Lemmas 1 and 2, we can show that every member of the set has a prome divisor and can be written as a product of promes. (You are invited to inspect the proofs of Lemmas 1 and 2 to see if any words need to be changed.) But an example shows that the prome decomposition of an integer in the set is not always unique:

$$693 = 21 \cdot 33 = 9 \cdot 77,$$

and 9, 21, 33, and 77 are all prome.

From the unique factorization theorem it follows that each positive integer can be written in exactly one way in the form

$$n = p_1^{e_1} p_2^{e_2} \cdots p_k^{e_k},$$

where $e_i \geq 1$, $i = 1, 2, \ldots, k$, each p_i is a prime, and $p_i \neq p_j$ for $i \neq j$. We call this representation the **prime-power decomposition** of n, and whenever we write

$$n = p_1^{e_1} p_2^{e_2} \cdots p_k^{e_k},$$

it will be understood, unless specified otherwise, that all the exponents are positive and the primes are distinct. The factor table in Appendix C gives the smallest prime that divides n for all n less than 10,000 and not divisible by 2 or 5. With the aid of this table, the prime-power decomposition for any $n \leq 10,000$ can be found readily. For example, take 8001. It is clearly not divisible by 2 or 5, and Table A gives its smallest prime factor as 3. Then $8001/3 = 2667$, and the table shows that 3 is a factor of 2667: $2667/3 = 889$. Again referring to the table, we see that $7 \mid 889$. Finally, $889/7 = 127$, which is prime. Thus

$$8001 = 3^2 \cdot 7 \cdot 127.$$

* **Exercise 5.** What is the prime-power decomposition of 7950?

To conclude this section, we note that the prime decomposition of integers gives another way of finding greatest common divisors besides the Euclidean algorithm. For example, consider $n = 120 = 2^3 \cdot 3 \cdot 5$ and $m = 252 = 2^2 \cdot 3^2 \cdot 7$. We see that 2^2 divides m and n, but no higher power of 2 is a common divisor of m and n. Also, 3 divides m and n, and no higher power of 3 is a common divisor. Furthermore, no other prime divides both m and n. Thus $2^2 \cdot 3$ is the greatest common divisor of m and n. Given the prime-power decompositions of m and n, we can write m and n as products of the same primes by inserting primes with the exponent zero where necessary. For example,

$$120 = 2^3 \cdot 3^1 \cdot 5^1 \cdot 7^0 \qquad \text{and} \qquad 252 = 2^2 \cdot 3^2 \cdot 5^0 \cdot 7^1.$$

In general, we have

Theorem 3. If $e_i \geq 0, f_i \geq 0$, $(i = 1, 2, \ldots, k)$,

$$m = p_1^{e_1} p_2^{e_2} \cdots p_k^{e_k} \qquad \text{and} \qquad n = p_1^{f_1} p_2^{f_2} \cdots p_k^{f_k};$$

then

$$(m, n) = p_1{}^{g_1}p_2{}^{g_2} \cdots p_k{}^{g_k},$$

where $g_i = \min(e_i, f_i)$, $i = 1, 2, \ldots, k$.

We will omit a formal proof, but you should have no trouble convincing yourself that it is true.

Problems

* 1. Find the prime-power decompositions of 1234, 34560, and 111111.

2. Find the prime-power decompositions of 2345, 45670, and 999999999999. (Note that $101 \mid 1000001$.)

3. Tartaglia (1556) claimed that the sums

$$1 + 2 + 4, \qquad 1 + 2 + 4 + 8, \qquad 1 + 2 + 4 + 8 + 16, \cdots$$

are alternately prime and composite. Show that he was wrong.

4. (a) DeBouvelles (1509) claimed that one or both of $6n + 1$ and $6n - 1$ are primes for all $n \geq 1$. Show that he was wrong.
 (b) Show that there are infinitely many n such that both $6n - 1$ and $6n + 1$ are composite.

5. Prove that if n is a square, then each exponent in its prime-power decomposition is even.

6. Prove that if each exponent in the prime-power decomposition of n is even, then n is a square.

*† 7. Find the smallest integer divisible by 2 and 3 which is simultaneously a square and a fifth power.

8. If $d \mid ab$, does it follow that $d \mid a$ or $d \mid b$?

* 9. Is it possible for a prime p to divide both n and $n + 1$ ($n \geq 1$)?

10. Prove that $n(n + 1)$ is never a square for $n > 0$.

*† 11. (a) Verify that $2^5 \cdot 9^2 = 2592$.
 (b) Is $2^5 \cdot a^b = 25ab$ possible for other a, b? (Here $25ab$ denotes the digits of $2^5 \cdot a^b$ and not a product.)

12. Let p be the least prime factor of n, where n is composite. Prove that if $p > n^{1/3}$, then n/p is prime.

*† 13. True or false? If p and q divide n, and each is greater than $n^{1/4}$, then n/pq is prime.

14. Prove that if n is composite, then 2^{n-1} is composite.

*† 15. Is it true that if $2^n - 1$ is composite, then n is composite?

Linear Diophantine Equations

Consider the following variation on an old problem:

> In a corral there are cowboys and an odd number of horses.
> There are 20 legs in all: how many belong to horses?

If we let h be the number of horses and c the number of cowboys in the corral, then we know that

(1) $$4h + 2c = 20,$$

assuming that all the horses and cowboys are whole. This equation has infinitely many solutions—for example,

h	-1	0	$5/3$	$\sqrt{2}$
c	12	10	$20/3$	$10 - 2\sqrt{2}$

But none of these fit the requirements of the problem: we want h and c to be *integers,* and positive ones at that.

Equations of this sort, in which we look for solutions in a restricted class of numbers—be they positive integers, negative integers, rational numbers, or whatever—are called **diophantine equations,** after Diophantus. Diophantus was probably an Egyptian who received a Greek education, long ago in Alexandria. It is not certain how long ago, but the difference between 1700 and 1850 years is not really important. What is important are the ideas he had, never had before by anyone

else, and the influence of those ideas. He started the evolution of our algebraic notation, and he was the first to pose and solve problems that called for solutions in integers or rational numbers. Other diophantine equations we will consider in later sections include

$$x^2 + y^2 = z^2, \qquad x^2 - 2y^2 = 1, \qquad \text{and} \qquad x^4 + y^4 = z^4,$$

where we will look for solutions in integers. All of these equations have infinitely many solutions in real or complex numbers, but the third has no solutions in integers except the trivial ones where either x or y is zero. In contrast, the first and second equations both have infinitely many solutions in integers.

In this section we will consider the simplest diophantine equation: the *linear diophantine equation*

$$ax + by = c,$$

where a, b, and c are integers. We want to find solutions in integers x and y. The equation $ax + by = c$ clearly has infinitely many solutions in rational numbers (and hence infinitely many solutions in real numbers), namely those given by

$$x = t, \qquad y = (c - at)/b$$

for any rational number t, if $b \neq 0$. But such an equation may have no solutions at all in integers. For example, $2x + 4y = 5$ has none.

* **Exercise 1.** Why not?

With the aid of results from Section 1, we can find all of the integer solutions of $ax + by = c$. Before we start, let us solve the horses and cowboys problem (1) by trial. Dividing both sides of the equation by 2, we have $2h + c = 10$, or

$$h = \frac{10 - c}{2}.$$

Since h and c must be positive integers, we may let $c = 1, 2, \ldots, 9$ (if $c > 9$, then h is not positive) and calculate the corresponding values of h:

c	1	2	3	4	5	6	7	8	9
h	9/2	4	7/2	3	5/2	2	3/2	1	1/2

Hence the diophantine equation has four solutions in positive integers: $(c, h) = (2, 4), (4, 3), (6, 2),$ and $(8, 1)$. But since the problem said that

the corral contained an odd number of horses and cowboys (plural), we get the unique answer: 12 legs belong to horses. Trial is sometimes the best way to solve a diophantine equation, but we want something surer.

If we can find just one solution of the linear diophantine equation, then we can find infinitely many. (In keeping with our convention that lower case italic letters denote integers unless we say otherwise, by ''solution'' we mean ''solution in integers.'') We prove this in

Lemma 1. If x_0, y_0 is a solution of $ax + by = c$, then so is

$$x_0 + bt, \qquad y_0 - at$$

for any integer t.

Proof. We are given that $ax_0 + by_0 = c$. Thus

$$a(x_0 + bt) + b(y_0 - at) = ax_0 + abt + by_0 - bat$$
$$= ax_0 + by_0$$
$$= c,$$

so $x_0 + bt, y_0 - at$ satisfies the equation too. For example, we can see by inspection that

$$5x + 6y = 17$$

is satisfied by $x = 1$ and $y = 2$. It follows from Lemma 1 that $x = 1 + 6t$, $y = 2 - 5t$ is also a solution for any integer t. Thus we can write down as many solutions as we please:

t	0	1	-1	3	-5	17	-1000
x	1	7	-5	19	-29	103	-5999
y	2	-3	7	-13	27	-83	5002

Each pair x, y satisfies $5x + 6y = 17$.

Exercise 2. Find by inspection a solution of $x + 5y = 10$ and use it to write five other solutions.

The next lemma lets us know when an equation has a solution and when it does not.

Lemma 2. If $(a, b) \nmid c$, then $ax + by = c$ has no solutions, and if $(a, b) \mid c$, then $ax + by = c$ has a solution.

Proof. Suppose that there are integers x_0, y_0 such that $ax_0 + by_0 = c$. Since $(a, b)|ax_0$ and $(a, b)|by_0$, it follows that $(a, b)|c$. Conversely, suppose that $(a, b)|c$. Then $c = m(a, b)$ for some m. From Theorem 4 of Section 1, we know that there are integers r and s such that

$$ar + bs = (a, b).$$

Then

$$a(rm) + b(sm) = m(a, b) = c,$$

and $x = rm$, $y = sm$ is a solution.

* **Exercise 3.** Which of the following linear diophantine equations is impossible? (We will say that a diophantine equation is *impossible* if it has no solutions.)

 (a) $14x + 34y = 90$.　　　(b) $14x + 35y = 91$.　　　(c) $14x + 36y = 93$.

Put $d = (a, b)$. Lemma 2 says that if $ax + by = c$ has a solution, then $d|c$. Put $a = da'$, $b = db'$, and $c = dc'$. If we divide $ax + by = c$ by d we get

$$a'x + b'y = c';$$

this equation has the same set of solutions as $ax + by = c$, and we know from Theorem 1 of Section 1 that $(a', b') = 1$. Thus, if a linear diophantine equation has solutions, then we can find them from an equation whose coefficients are relatively prime. For example, the first two equations of Exercise 3 are

$$7x + 17y = 45 \quad \text{and} \quad 2x + 5y = 13,$$

and $(7, 17) = (2, 5) = 1$.

The equation $2x + 5y = 13$ has for one solution $x = 4$ and $y = 1$, and from Lemma 1 we know that $x = 4 + 5t$, $y = 1 - 2t$ is a solution for any integer t. In the next lemma, we will show that these are *all* the solutions to the equation. The problem of finding all solutions of a diophantine equation is quite distinct from the problem of finding some solutions. It is also more difficult in general. For example, the equation

$$x^3 + y^3 = z^3 + w^3$$

has solutions given by

$$x = 1 - (s - 3t)(s^2 + 3t^2),$$
$$y = -1 + (s + 3t)(s^2 + 3t^2),$$

$$z = s + 3t - (s^2 + 3t^2)^2,$$
$$w = -s + 3t + (s^2 + 3t^2)^2,$$

where s and t may be any integers. You may verify this by multiplication, if you have the patience, but not all integer solutions are given by this formula.

Lemma 3. Suppose that $(a, b) = 1$ and x_0, y_0 is a solution of $ax + by = c$. Then all solutions of $ax + by = c$ are given by

$$x = x_0 + bt,$$
$$y = y_0 - at,$$

where t is an integer.

Proof. We see from Lemma 2 that the equation does have a solution, because $(a, b) = 1$ and $1 \mid c$ for all c. Then, let r, s be *any* solution of $ax + by = c$. We want to show that $r = x_0 + bt$ and $s = y_0 - at$ for some integer t. From $ax_0 + by_0 = c$ follows

$$c - c = (ax_0 + by_0) - (ar + bs)$$

or

(2) $$a(x_0 - r) + b(y_0 - s) = 0.$$

Because $a \mid a(x_0 - r)$ and $a \mid 0$, we have $a \mid b(y_0 - s)$. But we have supposed that a and b are relatively prime. It follows from Corollary 1 of Section 1 that $a \mid (y_0 - s)$. That is, there is an integer t such that

(3) $$at = y_0 - s.$$

Substituting in (2), this gives

$$a(x_0 - r) + bat = 0;$$

because $a \neq 0$, we may cancel it to get

(4) $$x_0 - r + bt = 0.$$

But (3) and (4) say that

$$s = y_0 - at,$$
$$r = x_0 + bt;$$

since r, s was *any* solution, the lemma is proved.

For example, $19x + 80y = 1980$ has a solution (found by inspection) of $x = 100$ and $y = 1$. Lemma 3 then says that all solutions are given by

$$x = 100 + 80t, \qquad y = 1 - 19t,$$

where t is an integer. In particular, all solutions with x and y positive are $(x, y) = (100, 1)$ and $(20, 20)$.

Up to now, we have assumed that neither a nor b was 0. If either is 0, the problem of solving $ax + by = c$ is trivial. If $a = 0$, then x can take on any value, and y can take on one or more, depending on whether $by = c$ has or does not have a solution in integers. It is similar if $b = 0$.

We can summarize the results of Lemmas 1 to 3 as follows:

Theorem 1. The linear diophantine equation $ax + by = c$ has no solutions if $(a, b) \nmid c$. If $(a, b) \mid c$, there are infinitely many solutions,

$$x = r + \frac{b}{(a, b)} t, \qquad y = s - \frac{a}{(a, b)} t,$$

where r, s is any solution and t is an integer.

In Section 5 we will see how to find the solution r, s using congruences. Theorem 1 should not be committed to memory, since it is only a statement of the process of solving linear diophantine equations, and the process is easy. For example, let us find all the solutions of $2x + 6y = 18$. Dividing out the common factor, we have $x + 3y = 9$. By inspection, $y = 0$, $x = 9$ is a solution. Hence all solutions are given by

(5) $$x = 9 + 3t, \qquad y = -t,$$

where t is an integer.

* **Exercise 4.** Find all solutions of $2x + 6y = 20$.

* **Exercise 5.** Find all the solutions of $2x + 6y = 18$ in *positive* integers. (Note that from (5), this is the same as asking for integers t such that $9 + 3t > 0$ and $-t > 0$.)

Linear diophantine equations can often be disguised in the "story problem" so dreaded by students, such as the one at the start of this section. Several of them appear among the problems, and they are not there solely to make students' lives miserable. Solving one is doing in miniature what a mathematician must do when confronted with a real problem from the real world: translate the problem into a mathematical one (that is the hard part), deal with the mathematical problem (that can be hard too), and apply the result to the original problem.

Problems

* 1. Find all the integer solutions of $x + y = 2$, $3x - 4y = 5$, and $15x + 16y = 17$.

 2. Find all the integer solutions of $2x + y = 2$, $3x - 4y = 0$, and $15x + 18y = 17$.

* 3. Find the solutions in positive integers of $x + y = 2$, $3x - 4y = 5$, and $6x + 15y = 51$.

 4. Find all the solutions in positive integers of $2x + y = 2$, $3x - 4y = 0$, and $7x + 15y = 51$.

* 5. Find all the positive solutions in integers of

$$x + y + z = 31,$$
$$x + 2y + 3z = 41.$$

 6. Find the five different ways a collection of 100 coins—pennies, dimes, and quarters—can be worth exactly $4.99.

* 7. A man bought a dozen pieces of fruit—apples and oranges—for 99 cents. If an apple costs 3 cents more than an orange, and he bought more apples than oranges, how many of each did he buy?

 8. The enrollment in a number theory class consists of sophomores, juniors, and backward seniors. If each sophomore contributes $1.25, each junior $.90, and each senior $.50, the instructor will have a fund of $25. There are 26 students; how many of each?

* 9. The following problem first appeared in an Indian book written around 850 AD. Three merchants found a purse along the way. One of them said, "If I secure this purse, I shall become twice as rich as both of you with your money on hand." Then the second said, "I shall become thrice as rich as both of you." The third man said, "I shall become five times as rich as both of you." How much did each merchant have, and how much was in the purse?

 10. A man cashes a check for d dollars and c cents at a bank. Assume that the teller by mistake gives the man c dollars and d cents. Assume that the man does not notice the error until he has spent 23 cents. Assume further that he then notices that he has $2d$ dollars and $2c$ cents. Assume still further that he asks you what amount the check was for. Assuming that you can accept all the assumptions, what is the answer?

Section

Congruences

Besides being quite pretty, congruences have many applications and will be used constantly in what follows. No one who lacks an acquaintance with congruences can claim to know much about number theory. As an example of their usefulness, it is easy to show, by using congruences, that no integer of the form $8n + 7$ is a sum of three squares. We will verify this later.

We say that *a is congruent to b modulo m* (in symbols, $a \equiv b \pmod{m}$) if and only if $m \mid (a - b)$, and we will suppose always that $m > 0$.

For example, $1 \equiv 5 \pmod{4}$, $-2 \equiv 9 \pmod{11}$, $6 \equiv 20 \pmod{7}$, and $720 \equiv 0 \pmod{10}$.

* **Exercise 1.** True or false? $91 \equiv 0 \pmod{7}$. $3 + 5 + 7 \equiv 5 \pmod{10}$. $-2 \equiv 2 \pmod{8}$. $11^2 \equiv 1 \pmod{3}$.

Really, $m \mid (a - b)$ and $a \equiv b \pmod{m}$ are only different notations for the same property, but a good notation can make things easy to see. Notation is vital. The ancient Greek mathematicians from 600 BC to 300 AD did not develop algebra at all, though they did such a fine job with geometry that Euclid's *Elements* was used as a textbook for 2000 years. It was not because they could not have—there is no question that Archimedes could have solved the general cubic polynomial equation, perhaps in his head, if he had put his mind to it *and* had had a satisfac-

27

tory notation—it was the lack of notation that stopped them. The solution of the cubic had to wait another 1700 years or so. The congruence notation was invented by the great mathematician and physicist Carl Friedrich Gauss (1777–1855), to whom is due quite a bit of the content of this book, and any other number theory book. His notation simplifies the proofs of theorems that would be difficult even to state without it. Another example of the value of a good notation is in calculus, in which Leibniz's notation (dy/dx, $\int y\ dx$) was superior to Newton's (\dot{y}, y').

There is another way to look at congruences:

Theorem 1. $a \equiv b$ (mod m) if and only if there is an integer k such that $a = b + km$.

Proof. Suppose that $a \equiv b$ (mod m). Then, from the definition of congruence, $m \mid (a - b)$. From the definition of divisibility, we know that since there is an integer k such that $km = a - b$, then $a = b + km$. Conversely, suppose that $a = b + km$.

* **Exercise 2.** Complete the proof.

Theorem 2. Every integer is congruent (mod m) to exactly one of 0, 1, . . . , $m - 1$.

Proof. Write $a = qm + r$, with $0 \le r < m$. We know from Theorem 2 of Section 1 that q and r are uniquely determined. Since $a \equiv r$ (mod m), the theorem is proved.

We call the number r in the last theorem the **least residue** of a (mod m). For example, the least residues of 71 modulo 2, 3, 5, 7, and 11 are 1, 2, 1, 1, and 5, respectively.

* **Exercise 3.** To what least residue (mod 11) is each of 23, 29, 31, 37, and 41 congruent?

Yet another way of looking at congruences is given by

Theorem 3. $a \equiv b$ (mod m) if and only if a and b leave the same remainder on division by m.

Proof. If a and b leave the same remainder r when divided by m, then

$$a = q_1m + r \quad \text{and} \quad b = q_2m + r$$

for some integers q_1 and q_2. It follows that

$$a - b = (q_1m + r) - (q_2m + r) = m(l_1 - q_2).$$

From the definition of divisibility, we have $m \mid (a - b)$. From the definition of congruence, we conclude that $a \equiv b$ (mod m). To prove the converse, suppose that $a \equiv b$ (mod m). Then $a \equiv b \equiv r$ (mod m), where r is a least residue modulo m. Then from Theorem 1,

$$a = q_1m + r \quad \text{and} \quad b = q_2m + r$$

for some integers q_1 and q_2; since $0 \leq r < m$, the theorem is proved.

For example, divide 1609 by 197: the quotient is 8 and the remainder is 33. Divide 1215 by 197: the quotient is 6 and the remainder is 33. It follows that $1609 \equiv 1215$ (mod 197), and in fact

$$1609 - 1215 = 394 = 2 \cdot 197.$$

It follows from Theorems 1 and 3 that the phrases "$n \equiv 7$ (mod 8)," "$n = 7 + 8k$ for some integer k," and "n leaves the remainder 7 when divided by 8" are different ways of saying the same thing.

* **Exercise 4.** Say "n is odd" in three other ways.

Exercise 5. Prove that $p \mid a$ if and only if $a \equiv 0$ (mod p).

Congruence acts like equality in many ways.

Lemma 1. For integers a, b, c, and d
 (a) $a \equiv a$ (mod m).
 (b) If $a \equiv b$ (mod m), then $b \equiv a$ (mod m).
 (c) If $a \equiv b$ (mod m) and $b \equiv c$ (mod m), then $a \equiv c$ (mod m).
 (d) If $a \equiv b$ (mod m) and $c \equiv d$ (mod m), then $a + c \equiv b + d$ (mod m).
 (e) If $a \equiv b$ (mod m) and $c \equiv d$ (mod m), then $ac \equiv bd$ (mod m).

Proof. All of these follow directly from the definition of congruence. Here is a proof of (e): We are given that $b - a = km$ and $d - c = jm$ for some integers k and j; thus

$$ac - bd = ac - (a + km)(c + jm)$$
$$= ac - ac - ajm - ckm - kjm^2$$
$$= m(-aj - ck - kjm);$$

from the definition of congruence, $ac \equiv bd \pmod m$.

Exercises 6, 7, 8, and 9. Prove parts (a), (b), (c), and (d).

Note that the lemma implies that we may substitute in congruences just as we do in equations. For example, if $x \equiv 2 \pmod 5$, then

$$2x^2 - x + 3 \equiv 2 \cdot 4 - 2 + 3 \equiv 9 \equiv 4 \pmod 5.$$

Although $ab = ac$ and $a \neq 0$ imply $b = c$ for all integers a, b, and c, it is not true that $ab \equiv ac \pmod m$ and $a \not\equiv 0 \pmod m$ imply $b \equiv c \pmod m$. (The symbol $\not\equiv$ means "not congruent to.") For example,

$$3 \cdot 4 \equiv 2 \cdot 8 \pmod{12} \qquad \text{but} \qquad 4 \not\equiv 8 \pmod{12}.$$

* **Exercise 10.** Construct a like example for modulus 10.

Although we cannot cancel freely, all is not lost, as we shall see from

Theorem 4. If $ac \equiv bc \pmod m$ and $(c, m) = 1$, then $a \equiv b \pmod m$.

Proof. From the definition of congruence, $m \,|\, (ac - bc)$; consequently, $m \,|\, c(a - b)$. Because $(m, c) = 1$, we can conclude from Theorem 5 of Section 1 that $m \,|\, (a - b)$. That is, $a \equiv b \pmod m$.

* **Exercise 11.** What values of x satisfy

(a) $2x \equiv 4 \pmod 7$? \qquad (b) $2x \equiv 1 \pmod 7$?

(Hint for (b): $1 \equiv 8 \pmod 7$.)

We can, then, cancel a factor that appears on both sides of a congruence if it is relatively prime to the modulus. We now consider the case in which the factor and the modulus are not relatively prime.

Theorem 5. If $ac \equiv bc \pmod m$ and $(c, m) = d$, then $a \equiv b \pmod{m/d}$.

Proof. If $ac \equiv bc \pmod{m}$, then $m \,|\, c(a - b)$ and $m/d \,|\, (c/d)(a - b)$. Since we know that $(m/d, c/d) = 1$, from Theorem 5 of Section 1 we get $m/d \,|\, (a - b)$, so $a \equiv b \pmod{m/d}$.

That is, we can cancel a common factor from both sides of a congruence if we divide the modulus by its greatest common divisor with the factor. For example, $30x \equiv 27 \pmod{33}$ implies $10x \equiv 9 \pmod{11}$.

* **Exercise 12.** Which x will satisfy $2x \equiv 4 \pmod 6$?

Now we can see how easy it is to show that no integer of the form $8n + 7$ is the sum of three squares. Suppose that $k = 8n + 7$ is the sum of three squares. Then $k \equiv 7 \pmod 8$ and $k = a^2 + b^2 + c^2$ for some integers a, b, and c. Thus

$$a^2 + b^2 + c^2 \equiv 7 \pmod 8.$$

We now show that this last congruence is impossible for any integers a, b, and c. What values can x^2 assume, modulo 8? Every integer has one of 0, 1, 2, 3, 4, 5, 6, and 7 for a least residue (mod 8), and

$$0^2 \equiv 0, \qquad 1^2 \equiv 1, \qquad 2^2 \equiv 4, \qquad 3^2 \equiv 1$$
$$4^2 \equiv 0, \qquad 5^2 \equiv 1, \qquad 6^2 \equiv 4, \qquad 7^2 \equiv 1,$$

all modulo 8. Thus the square of any integer is congruent modulo 8 to one of 0, 1, and 4. It is impossible to make any combination of three numbers selected from 0, 1, and 4 add up to anything congruent to 7 (mod 8). (The statements $1 + 1 + 4 \equiv 6$ and $0 + 4 + 4 \equiv 8 \pmod 8$ are as close as you can come.) Hence $a^2 + b^2 + c^2$ is never congruent to 7 (mod 8) for any integers a, b, and c. Thus $a^2 + b^2 + c^2 = 8n + 7$ is an impossible equation.

You may already know the handy test for divisibility by 9: an integer is divisible by 9 if and only if the sum of its digits is divisible by 9, and you may already know why it is true. Congruences make the proof trivial. Since $10 \equiv 1 \pmod 9$, $10^n \equiv 1^n \equiv 1 \pmod 9$, and that is all we need to prove

Theorem 6. Every integer is congruent (mod 9) to the sum of its digits.

Proof. Take an integer n, and let its digital representation by

$$d_k d_{k-1} d_{k-2} \cdots d_1 d_0.$$

That is,

$$n = d_k 10^k + d_{k-1} 10^{k-1} + d_{k-2} 10^{k-2} + \cdots + d_1 10^1 + d_0 10^0$$
$$\equiv d_k + d_{k-1} + d_{k-2} + \cdots + d_0 \ (\text{mod } 9),$$

which is what we wanted to show.

This theorem shows why the process of casting out nines to check an addition or a multiplication works. The rule was a feature of many arithmetic books in the past, when long columns of numbers had to be added by hand and there were no mechanical devices to perform multiplications. Now we can all have pocket calculators, but it is wise to be prepared for the time when our batteries fail. If two numbers are equal, they are congruent to any modulus, 9 in particular. So if we are told that $(314)(159) = 49826$, we can see right away that we are being lied to, because

$$(314)(159) \equiv (3 + 1 + 4)(1 + 5 + 9) \equiv 8 \cdot 15 \equiv 8(1 + 5)$$
$$\equiv 48 \equiv 4 + 8 \equiv 12 \equiv 1 + 2 \equiv 3 \ (\text{mod } 9),$$

while

$$49826 \equiv 4 + 9 + 8 + 2 + 6 \equiv 29 \equiv 2 + 9 \equiv 11 \equiv 1 + 1$$
$$\equiv 2 \ (\text{mod } 9):$$

the numbers are not congruent (mod 9), so they cannot be equal.

Problems

* 1. Find the least residue of 1492 (mod 4), (mod 10), and (mod 101).
 2. Find the least residue of 1789 (mod 4), (mod 10), and (mod 101).
 3. Prove or disprove that if $a \equiv b \ (\text{mod } m)$, then $a^2 \equiv b^2 \ (\text{mod } m)$.
 4. Prove or disprove that if $a^2 \equiv b^2 \ (\text{mod } m)$, then $a \equiv b$ or $-b \ (\text{mod } m)$.
* 5. Find all m such that $1066 \equiv 1776 \ (\text{mod } m)$.
 6. Find all m such that $1848 \equiv 1914 \ (\text{mod } m)$.
* 7. If $k \equiv 1 \ (\text{mod } 4)$, then what is $6k + 5$ congruent to (mod 4)?
 8. Show that every prime (except 2) is congruent to 1 or 3 (mod 4).
 9. Show that every prime (except 2 or 3) is congruent to 1 or 5 (mod 6).
 10. What can primes (except 2, 3, or 5) be congruent to (mod 30)?

* 11. In the multiplication $31415 \cdot 92653 = 2910\ 93995$, one digit in the product is missing and all the others are correct. Find the missing digit without doing the multiplication.

12. Show that no square has as its last digit, 2, 3, 7, or 8.

* 13. What can the last digit of a fourth power be?

14. Show that the difference of two consecutive cubes is never divisible by 3.

15. Show that the difference of two consecutive cubes is never divisible by 5.

16. Show that

$$d_k 10^k + d_{k-1} 10^{k-1} + \cdots + d_1 10 + d_0$$
$$\equiv d_0 - d_1 + d_2 - d_3 + \cdots + (-1)^k d_k \pmod{11}$$

and deduce a test for divisibility by 11.

* 17. A says, "27,182,818,284,590,452 is divisible by 11." B says, "No, it isn't." Who is right?

18. A *palindrome* is a number that reads the same backward as forward. Examples are 22, 1331, and 935686539.

 (a) Prove that every four-digit palindrome is divisible by 11.
 (b) What about six-digit palindromes?

19. Show that if $n \equiv 4 \pmod 9$, then n cannot be written as the sum of three cubes.

20. Show that for $k > 0$ and $m \geq 1$, $x \equiv 1 \pmod{m^k}$ implies $x^m \equiv 1 \pmod{m^{k+1}}$.

Linear Congruences

After defining congruences and studying some of their properties, it is natural to look at congruences involving unknowns, like $3x \equiv 4 \pmod 5$ and $x^{17} + 3x - 3 \equiv 0 \pmod{31}$, and see how to solve them, if we can. The simplest such congruence is the **linear congruence** $ax \equiv b \pmod m$, and this is what this section is devoted to. The congruence $ax \equiv b \pmod m$ has a solution if and only if there are integers x and k such that $ax = b + km$. Hence, the problem of solving linear congruences is essentially the same as that of solving linear diophantine equations, and Theorem 1 of this section is the same as Theorem 1 of Section 3, but in a different notation.

If one integer satisfies $ax \equiv b \pmod m$, then there are infinitely many. For example, the table below shows that $3x \equiv 4 \pmod 5$ is true if $x = 3$

x	0	1	2	3	4	5	6	7	8	9
$3x \pmod 5$	0	3	1	4	2	0	3	1	4	2

or $x = 8$, and it is clear that it is also true if $x = 13, 18, 23, \ldots$ or $x = -2, -7, -12, \ldots$. In general, if r is an integer such that $ar \equiv b \pmod m$, then all of the integers $r + m, r + 2m, \ldots, r - m, r - 2m, \ldots$ satisfy the congruence, since

$$a(r + km) \equiv ar \equiv b \pmod m$$

for any integer k. Among the integers $r + km$, $k = 0$, ± 1, ± 2, . . . , there will be exactly one—say s—that satisfies $0 \leq s < m$. This is because every integer lies between two successive multiples of m. If r satisfies the congruence and $km \leq r < (k + 1)m$ for some k, then $0 \leq r - km < m$; we can put $s = r - km$. We will single this integer out and say that by a **solution** to $ax \equiv b$ (mod m), we mean a number r such that $ar \equiv b$ (mod m) and r is a least residue (mod m). Thus, *the* solution to $3x \equiv 4$ (mod 5) is 3, because it makes the congruence true and it is a least residue (mod 5).

Unlike the familiar linear equation $ax = b$, the linear congruence $ax \equiv b$ (mod m) may have no solutions, exactly one solution, or many solutions. For example, $2x \equiv 1$ (mod 3) is satisfied by $x = 2$ and for no other values of x that are least residues (mod 3). Hence it has just one solution, namely 2. The congruence $2x \equiv 1$ (mod 4) has no solutions, because $2x$ is congruent to 0 or 2 (mod 4) for any x. The congruence $2x \equiv 4$ (mod 6) has two solutions, 2 and 5.

* **Exercise 1.** Construct congruences modulo 12 with no solutions, just one solution, and more than one solution.

* **Exercise 2.** Which congruences have no solutions?

 (a) $3x \equiv 1$ (mod 10), (b) $4x \equiv 1$ (mod 10), (c) $5x \equiv 1$ (mod 10),
 (d) $6x \equiv 1$ (mod 10), (e) $7x \equiv 1$ (mod 10).

* **Exercise 3** (optional). After Exercise 2, can you guess a criterion for telling when a congruence has no solutions?

We will now set out to prove a theorem that will let us see how many solutions a linear congruence has.

Lemma 1. If $(a, m) \nmid b$, then $ax \equiv b$ (mod m) has no solutions.

Proof. We will prove the contrapositive, which is logically the same thing: if $ax \equiv b$ (mod m) has a solution, then $(a, m) \mid b$. Suppose that r is a solution. Then $ar \equiv b$ (mod m), and from the definition of congruence, $m \mid (ar - b)$, or from the definition of divides, $ar - b = km$ for some k. Since $(a, m) \mid a$ and $(a, m) \mid km$, it follows that $(a, m) \mid b$.

For example, $6x \equiv 7$ (mod 8) has no solutions.

Lemma 2. If $(a, m) = 1$, then $ax \equiv b \pmod m$ has exactly one solution.

Proof. Since $(a, m) = 1$, we know that there are integers r and s such that $ar + ms = 1$. Multiplying by b gives

$$a(rb) + m(sb) = b.$$

We see that $arb - b$ is a multiple of m, or

$$a(rb) \equiv b \pmod m.$$

The least residue of rb modulo m is then a solution of the linear congruence.

It remains to show that there is not more than one solution. Suppose that both r and s are solutions. That is, since

$$ar \equiv b \pmod m \qquad \text{and} \qquad as \equiv b \pmod m,$$

then $ar \equiv as \pmod m$. Because $(a, m) = 1$, we can apply Theorem 4 of the last section, cancel the common factor, and get $r \equiv s \pmod m$. That is, $m \mid (r - s)$. But r and s are least residues $\pmod m$, so

$$0 \le r < m \qquad \text{and} \qquad 0 \le s < m.$$

Thus $-m < r - s < m$; together with $m \mid (r - s)$, this gives $r - s = 0$, or $r = s$, and the solution is unique. The above argument is quite general and will be used often: if two least residues $\pmod m$ are congruent $\pmod m$, then they are equal.

Inspection is one way of solving congruences with small moduli, and another is substituting all possible values for the variable. But the best way is to manipulate the coefficients until cancellation is possible. For example, to solve $4x \equiv 1 \pmod{15}$, we can write

$$4x \equiv 1 \equiv 16 \pmod{15}$$

and cancel to get $x \equiv 4 \pmod{15}$. As another example, let us solve $14x \equiv 27 \pmod{31}$. From

$$14x \equiv 27 \equiv 58 \pmod{31}$$

we get $7x \equiv 29 \pmod{31}$. We continue adding 31 until we can cancel:

$$7x \equiv 29 \equiv 60 \equiv 91 \pmod{31},$$

so we get $x \equiv 13 \pmod{31}$, and 13 is the solution.

This method is the best to use when solving linear diophantine equations. The equation $ax + by = c$ implies the two congruences

$$ax \equiv c \ (\text{mod } b) \qquad \text{and} \qquad by \equiv c \ (\text{mod } a).$$

We can choose either one, solve for the variable, and then substitute the result into the original equation to get all the solutions. For example, let us solve $9x + 16y = 35$. This gives $16y \equiv 35 \ (\text{mod } 9)$ or $7y \equiv 35$ $(\text{mod } 9)$, from which we get $y \equiv 5 \ (\text{mod } 9)$. That is, $y = 5 + 9t$ for some integer t. Substituting this in the original equation, we get

$$9x + 16(5 + 9t) = 35,$$

or $9x + 144t = -45$, or $x + 16t = -5$. We thus have all the solutions:

$$x = -5 - 16t,$$
$$y = 5 + 9t,$$

t an integer.

* **Exercise 4.** Solve

$$\text{(a) } 8x \equiv 1 \ (\text{mod } 15), \qquad \text{(b) } 9x + 10y = 11.$$

There remains the case where $(a, m) \mid b$ and $(a, m) \neq 1$. Cancellation reduces this to the case $(a, m) = 1$. For example, to solve $6x \equiv 15 \ (\text{mod } 33)$, apply Theorem 4 of the last section to get $2x \equiv 5 \ (\text{mod } 11)$ which is satisfied by all integers $x \equiv 8 \ (\text{mod } 11)$. The solutions to the original congruence are all of the least residues $(\text{mod } 33)$ which satisfy it, and these are 8, 19, and 30: $(6, 33) = 3$, and the congruence has three solutions. This is what happens in general.

Lemma 3. Let $d = (a, m)$. If $d \mid b$, then $ax \equiv b \ (\text{mod } m)$ has exactly d solutions.

Proof. If we cancel the common factor, we get a congruence $(a/d)x \equiv (b/d) \ (\text{mod } m/d)$, which we know has exactly one solution, because $(a/d, m/d) = 1$. Call it r, and let s be any other solution of $ax \equiv b \ (\text{mod } m)$. Then $ar \equiv as \equiv b \ (\text{mod } m)$, and it follows from Theorem 5 of the last section that $r \equiv s \ (\text{mod } m/d)$. That is, $s - r = k(m/d)$ or $s = r + k(m/d)$ for some k. Putting $k = 0, 1, \ldots, d - 1$, we get numbers which are least residues $(\text{mod } m)$, since

$$0 \leq r + k(m/d) < (m/d) + (d - 1)(m/d) = d(m/d) = m,$$

and they all satisfy $ax \equiv b$ (mod m), because

$$(a/d)(r + k(m/d)) \equiv (a/d)r \equiv b/d \pmod{m/d},$$

and this implies

$$a(r + k(m/d)) \equiv b \pmod{m}.$$

* **Exercise 5.** Determine the number of solutions of each of the following congruences:

$$3x \equiv 6 \text{ (mod 15)}, \qquad 4x \equiv 8 \text{ (mod 15)}, \qquad 5x \equiv 10 \text{ (mod 15)},$$
$$6x \equiv 11 \text{ (mod 15)}, \qquad 7x \equiv 14 \text{ (mod 15)}.$$

* **Exercise 6.** Find all of the solutions of $5x \equiv 10$ (mod 15).

We can summarize the results of Lemmas 1 to 3 in

Theorem 1. $ax \equiv b$ (mod m) has no solutions if $(a, m) \nmid b$. If $(a, m) \mid b$, then there are exactly (a, m) solutions.

* **Exercise 7.** Solve the rest of the congruences in Exercise 5.

In a Chinese work on mathematics written more than 1000 years ago, there was a problem like, "Find a number that leaves a remainder of 1 when divided by 3, of 2 when divided by 5, and of 3 when divided by 7." It is hard to imagine a practical situation where such a problem could arise. Mathematics was developed to deal with problems coming from commerce, government, astronomy, and religion, and this one does not seem to come from any one of those. Yet Babylonian clay tablets, written more than 2000 years ago, contain some types of cubic polynomial equations and their solutions, and those do not arise in everyday life either. Both are examples of the fascination that mathematics has had for the human mind—for some human minds, anyway. After the practical problems are solved, mathematics is interesting for its own sake.

We will now consider, for its own sake, and because it is useful elsewhere in number theory, how to solve problems like the ancient Chinese one mentioned above. In our notation, the problem is to find x such that

$$x \equiv 1 \text{ (mod 3)}, \qquad x \equiv 2 \text{ (mod 5)}, \qquad \text{and} \qquad x \equiv 3 \text{ (mod 7)}.$$

Exercise 8. Verify that 52 satisfies each of the three congruences.

The first congruence says that $x = 1 + 3k_1$ for some k_1. Substituting this into the second congruence, we see that k_1 must satisfy

$$1 + 3k_1 \equiv 2 \ (\text{mod } 5).$$

Solving, we get $k_1 \equiv 2 \ (\text{mod } 5)$. That is, $k_1 = 2 + 5k_2$ for some k_2, and thus

$$x = 1 + 3k_1 = 1 + 3(2 + 5k_2) = 7 + 15k_2.$$

This x satisfies the first two congruences. If in addition x satisfies the third, we must have

$$7 + 15k_2 \equiv 3 \ (\text{mod } 7).$$

Solving for k_2, we get $k_2 \equiv 3 \ (\text{mod } 7)$. Thus

$$x = 7 + 15(3 + 7k_3) = 52 + 105k_3$$

satisfies all three congruences for any integer k_3. Otherwise expressed, any $x \equiv 52 \ (\text{mod } 105)$ satisfies the three congruences. In fact, 52 is the unique solution modulo 105.

The next theorem amounts to nothing more than a statement that the same procedure will always work.

Theorem 2. The Chinese Remainder Theorem. The system of congruences

(1) $$x \equiv a_i \ (\textit{mod } m_i), \qquad i = 1, 2, \ldots, k,$$

where $(m_i, m_j) = 1$ if $i \neq j$, has a unique solution modulo $m_1 m_2 \cdots m_k$.

Proof. We first show, by induction, that system (1) has a solution. The result is obvious when $k = 1$. Let us consider the case $k = 2$. If $x \equiv a_1$ (mod m_1), then $x = a_1 + k_1 m_1$ for some k_1. If in addition $x \equiv a_2$ (mod m_2), then

$$a_1 + k_1 m_1 \equiv a_2 \ (\text{mod } m_2)$$

or

$$k_1 m_1 \equiv a_2 - a_1 \ (\text{mod } m_2).$$

Because $(m_2, m_1) = 1$, we know that this congruence, with k_1 as the unknown, has a unique solution modulo m_2. Call it t. Then $k_1 = t + k_2 m_2$ for some k_2, and

$$x = a_1 + (t + k_2m_2)m_1 \equiv a_1 + tm_1 \pmod{m_1m_2}$$

satisfies both congruences.

Suppose that system (1) has a solution (mod $m_1m_2 \cdots m_k$) for $k = r - 1$. Then there is a solution, s, to the system

$$x \equiv a_i \pmod{m_i}, \qquad i = 1, 2, \ldots, r - 1.$$

But the system

$$x \equiv s \pmod{m_1m_2 \cdots m_{r-1}},$$
$$x \equiv a_r \pmod{m_r}$$

has a solution modulo the product of the moduli, just as in the case $k = 2$, because $(m_1m_2 \cdots m_{k-1}, m_k) = 1$. (This statement is true because no prime that divides $m_i, i = 1, 2, \ldots, k - 1$, can divide m_k.)

It is easy to see that the solution is unique. If r and s are both solutions of the system, then

$$r \equiv s \equiv a_i \pmod{m_i}, \qquad i = 1, 2, \ldots, k,$$

so $m_i | (r - s), i = 1, 2, \ldots, k$. Thus $r - s$ is a common multiple of m_1, m_2, \ldots, m_k, and because the moduli are relatively prime in pairs, we have $m_1m_2 \cdots m_k | (r - s)$. But since r and s are least residues modulo $m_1m_2 \cdots m_k$,

$$-m_1m_2 \cdots m_k < r - s < m_1m_2 \cdots m_k,$$

whence $r - s = 0$.

Problems

*† 1. Solve each of the following:

$$2x \equiv 1 \pmod{17}. \qquad 3x \equiv 1 \pmod{17}.$$
$$3x \equiv 6 \pmod{18}. \qquad 40x \equiv 777 \pmod{1777}.$$

2. Solve each of the following:

$$2x \equiv 1 \pmod{19}. \qquad 3x \equiv 1 \pmod{19}.$$
$$4x \equiv 6 \pmod{18}. \qquad 20x \equiv 984 \pmod{1984}.$$

* 3. Solve the systems

(a) $x \equiv 1 \pmod{2}, x \equiv 1 \pmod{3}$.
(b) $x \equiv 3 \pmod{5}, x \equiv 5 \pmod{7}, x \equiv 7 \pmod{11}$.
(c) $2x \equiv 1 \pmod{5}, 3x \equiv 2 \pmod{7}, 4x \equiv 3 \pmod{11}$.

4. Solve the systems

(a) $x \equiv 1 \pmod{2}, x \equiv 2 \pmod{3}$.

(b) $x \equiv 2 \pmod 5$, $2x \equiv 3 \pmod 7$, $3x \equiv 4 \pmod{11}$.

(c) $x \equiv 31 \pmod{41}$, $x \equiv 59 \pmod{26}$.

* 5. What possibilities are there for the number of solutions of a linear congruence (mod 20)?

6. Construct linear congruences modulo 20 with no solutions, just one solution, and more than one solution. Can you find one with 20 solutions?

* 7. Solve $9x \equiv 4 \pmod{1453}$.

8. Solve $4x \equiv 9 \pmod{1453}$.

* 9. Solve for x and y:

(a) $x + 2y \equiv 3 \pmod 7$, $3x + y \equiv 2 \pmod 7$.

(b) $x + 2y \equiv 3 \pmod 6$, $3x + y \equiv 2 \pmod 6$.

10. Solve for x and y:

(a) $x + 2y \equiv 3 \pmod 9$, $3x + y \equiv 2 \pmod 9$.

(b) $x + 2y \equiv 3 \pmod{10}$, $3x + y \equiv 2 \pmod{10}$.

* 11. When the marchers in the annual Mathematics Department Parade lined up 4 abreast, there was 1 odd person; when they tried 5 in a line, there were 2 left over; and when 7 abreast, there were 3 left over. How large is the Department?

12. Find a multiple of 7 that leaves the remainder 1 when divided by 2, 3, 4, 5, or 6.

* 13. Find the smallest odd n, $n > 3$, such that $3 \mid n$, $5 \mid n + 2$, and $7 \mid n + 4$.

14. Find the smallest integer n, $n > 2$, such that $2 \mid n$, $3 \mid n + 1$, $4 \mid n + 2$, $5 \mid n + 3$, and $6 \mid n + 4$.

*† 15. Find a positive integer such that half of it is a square, a third of it is a cube, and a fifth of it is a fifth power.

16. The three consecutive integers 48, 49, and 50 each have a square factor.

(a) Find n such that $3^2 \mid n$, $4^2 \mid n + 1$, and $5^2 \mid n + 2$.

(b) Can you find n such that $2^2 \mid n$, $3^2 \mid n + 1$, and $4^2 \mid n + 2$?

17. If $x \equiv r \pmod m$ and $x \equiv s \pmod{m + 1}$, show that

$$x \equiv r(m + 1) - sm \pmod{m(m + 1)}.$$

18. What three positive integers, upon being multiplied by 3, 5, and 7 respectively and the products divided by 20, have remainders in arithmetic progression with common difference 1 and quotients equal to remainders?

*† 19. Suppose that the moduli in the system

$$x \equiv a_i \pmod{m_i}, \qquad i = 1, 2, \ldots, k$$

are not relatively prime in pairs. Find a condition that the a_i must satisfy in order that the system have a solution.

20. How many multiples of b are there in the sequence

$$a, 2a, 3a, \ldots, ba?$$

Section

6

Fermat's and Wilson's Theorems

In this section we will prove

Theorem 1. Fermat's Theorem. If p is prime and $(a, p) = 1$, then

$$a^{p-1} \equiv 1 \pmod{p}$$

This theorem was first stated, without proof, by Fermat in 1640. Fermat was a French lawyer and judge who did mathematics in his spare time and for the fun of it. Besides his large contributions to number theory, he was, with Pascal, responsible for the beginnings of probability theory. He was the most remarkable amateur in the history of mathematics, both in the original meaning of "lover of" and in the present-day meaning of "nonprofessional," and I think that he had the best mathematical mind of the seventeenth century, after Newton's. His theorem is vital, as we shall see in the study of quadratic congruences, and it has many other applications. Its statement and proof are simple, but its effects are great. We will also prove

Theorem 2. Wilson's Theorem. p is a prime if and only if

$$(p - 1)! \equiv -1 \pmod{p}.$$

Recall that $n! = n(n-1)(n-2) \ldots 3 \times 2 \times 1$ for $n \geq 1$ and $0! = 1$, by definition. Factorials grow rapidly; 2! is only 2 and 3! = 6, but 15! is already 1307674368000. Wilson's Theorem is not really Wilson's. He only guessed that it was true and told the mathematician Waring about it. Waring was unable to prove it either, but he published it without proof in 1770. It was almost immediately proved by Lagrange, one of the most powerful of eighteenth-century mathematicians. Wilson was not even the first to guess the theorem: Leibniz also noted it in 1682, perhaps while resting from developing calculus. However, Wilson's name is inextricably wedded to the theorem, and through it he has achieved mathematical immortality.

Wilson's Theorem is remarkable because it gives a condition both necessary and sufficient for a number to be prime. Thus, in theory, the problem of determining whether a given number is prime is completely solved. But for large integers, the computational difficulties are great. For the moderate-sized prime

$$p = 162{,}259{,}276{,}829{,}213{,}363{,}391{,}578{,}010{,}288{,}127,$$

the calculation of the least residue of $(p-1)! \pmod{p}$ would take about 10^{33} multiplications of two 33-digit numbers, each followed by division by p. Even our fastest computers are not fast enough. Compare, for example, the calculation of 12! (mod 13) with the labor in verifying that 13 is divisible by neither 2 nor 3.

To see that Fermat's Theorem is true in a special case, take $p = 7$. Successive multiplication gives (mod 7),

a	a^2	a^3	a^4	a^5	a^6
1	1	1	1	1	1
2	4	1	2	4	1
3	2	6	4	5	1
4	2	1	4	2	1
5	4	6	2	3	1
6	1	6	1	6	1

so $a^6 \equiv 1 \pmod{7}$ for all a such that $(a, 7) = 1$. Wilson's Theorem is also true for small values of n:

n	2	3	4	5	6	7	8	9	10	11
$(n-1)!$	1	2	6	24	120	720	5040	40320	362880	3628800
$(n-1)! \pmod{n}$	1	2	2	4	0	6	0	0	0	10

Note that both Fermat's and Wilson's Theorems are true when $p = 2$. As always, p denotes a prime, but in the rest of this section we will assume that p is an odd prime. That 2 is a prime is sometimes exasperating, but nothing can be done about it.

We start the proof of Fermat's Theorem with

Lemma 1. If $(a, m) = 1$, then the least residues of

(1) $a,\ 2a,\ 3a,\ \dots\ ,\ (m-1)a \pmod{m}$,

are

(2) $1,\ 2,\ 3,\ \dots\ ,\ m-1$

is some order.

Stated differently, if $(a, m) = 1$, then each integer is congruent (mod m) to exactly one of $a,\ 2a,\ \dots\ ,\ (m-1)a$. For example, take $m = 8$ and $a = 3$: the numbers in (1) are then

$$3,\ 6,\ 9,\ 12,\ 15,\ 18,\ 21,$$

and their least residues (mod 8) are

$$3,\ 6,\ 1,\ 4,\ 7,\ 2,\ 5.$$

Proof. There are $m - 1$ numbers in (1), none congruent to 0 (mod m). Hence each of them is congruent (mod m) to one of the numbers in (2). If we show that no two of the integers in (1) are congruent (mod m), then it follows that their least residues (mod m) are all different, and hence are a permutation of $1, 2, \dots, m - 1$. Suppose that two of the integers in (1) are congruent (mod m): that is,

$$ra \equiv sa \pmod{m};$$

because $(a, m) = 1$ we can cancel (Theorem 4 of Section 4) and get

$$r \equiv s \pmod{m}.$$

But r and s are least residues modulo m; by an argument we have used several times before, it follows that $r = s$. This proves the lemma.

Proof of Fermat's Theorem. Given any prime p, Lemma 1 says that if $(a, p) = 1$, then the least residues of

$$a,\ 2a,\ \dots\ ,\ (p-1)a \pmod{p}$$

are a permutation of

$$1, 2, \ldots, p - 1.$$

Hence their products are congruent (mod p):

$$a \cdot 2a \cdot 3a \cdots (p - 1)a \equiv 1 \cdot 2 \cdot 3 \cdots (p - 1) \pmod{p},$$

or

$$a^{p-1}(p - 1)! \equiv (p - 1)! \pmod{p}.$$

Since p and $(p - 1)!$ are relatively prime, the last congruence gives

$$a^{p-1} \equiv 1 \pmod{p},$$

which is Fermat's Theorem.

Exercise 1. Verify that the theorem is true for $a = 2$ and $p = 5$.

Fermat's Theorem is sometimes stated in a slightly different way:

Corollary. If p is a prime, then

$$a^p \equiv a \pmod{p}$$

for all a.

Proof. If $(a, p) = 1$, this follows from Fermat's Theorem. If $(a, p) = p$, then the corollary says $0 \equiv 0 \pmod{p}$, which is true. There are no other cases.

As an example, let us verify that $3^{16} \equiv 1 \pmod{17}$. It is not necessary to calculate the large integer 3^{16} and then divide it by 17; we can proceed in stages, reducing modulo 17 as we go. We have

$$3^3 \equiv 27 \equiv 10 \pmod{17}.$$

Squaring, we get

$$3^6 \equiv 100 \equiv -2 \pmod{17};$$

squaring again yields $3^{12} \equiv 4 \pmod{17}$. Thus

$$3^{16} \equiv 3^{12} \cdot 3^3 \cdot 3 \equiv 4 \cdot 10 \cdot 3 \equiv 120 \equiv 1 \pmod{17}.$$

* **Exercise 2.** Calculate 2^2 and 20^{10} (mod 11).

To prove Wilson's Theorem, we need two lemmas:

Lemma 2. $x^2 \equiv 1 \pmod p$ has exactly two solutions: 1 and $p - 1$.

Proof. Let r be any solution of $x^2 \equiv 1 \pmod p$. By *solution* we mean, as we did for linear congruences, a least residue that satisfies the congruence. We have $r^2 - 1 \equiv 0 \pmod p$, so

$$p \,|\, (r + 1)(r - 1).$$

Hence $p \,|\, (r + 1)$ or $p \,|\, (r - 1)$; otherwise expressed,

$$r + 1 \equiv 0 \quad \text{or} \quad r - 1 \equiv 0 \pmod p,$$

so $r \equiv p - 1$ or 1 (mod p). Since r is a least residue (mod p), it follows that $r = p - 1$ or 1. It is easy to verify that both of these numbers actually satisfy $x^2 \equiv 1 \pmod p$.

Lemma 2 has a familiar analogy: $x^2 = 1$ is satisfied only when $x = 1$ or $x = -1$, and $x^2 \equiv 1 \pmod p$ is satisfied only when $x \equiv 1 \pmod p$ or $x \equiv -1 \pmod p$.

If $(a, p) = 1$, we know that $ax \equiv 1 \pmod p$ has exactly one solution. Let us denote it by a'; it is for congruence what the reciprocal is for equality, since $aa' \equiv 1 \pmod p$. For example, if $p = 13$ we have

a	1	2	3	4	5	6	7	8	9	10	11	12
a'	1	7	9	10	8	11	2	5	3	4	6	12

Note that there no duplications in the second line. This is no accident, and it is true in general; this is the content of the next lemma.

Lemma 3. Let p be an odd prime and let a' be the solution of $ax \equiv 1$ (mod p), $a = 1, 2, \ldots, p - 1$. $a' \equiv b' \pmod p$ if and only if $a \equiv b$ (mod p). Furthermore, $a \equiv a' \pmod p$ if and only if $a = 1$ or $p - 1$.

Proof. Suppose that $a' \equiv b' \pmod p$. Then

$$b \equiv aa'b \equiv ab'b \equiv a \pmod p.$$

Conversely, suppose that $a \equiv b \pmod{p}$. Then

$$b' \equiv b'aa' \equiv b'ba' \equiv a' \pmod{p}.$$

For the second part of the proof, it follows from $1 \cdot 1 \equiv (p-1)$ $(p-1) \equiv 1 \pmod{p}$ that $1' \equiv 1 \pmod{p}$ and $(p-1)' \equiv p-1 \pmod{p}$. Conversely, if $a \equiv a' \pmod{p}$, then $1 \equiv aa' \equiv a^2 \pmod{p}$, and from Lemma 2 we know that this is possible only if $a = 1$ or $p - 1$.

Proof of Wilson's Theorem. From Lemma 3, we know that we can separate the numbers

$$2, 3, \ldots, p - 2$$

into $(p-3)/2$ pairs such that each pair consists of an integer a and its associated a', which is different from a. For example, for $p = 13$ the pairs

$$(2, 7), (3, 9), (4, 10), (5, 8), (6, 11).$$

* **Exercise 3.** What are the pairs when $p = 11$?

The product of the two integers in each pair is congruent to 1 (mod p), so it follows that

$$2 \cdot 3 \cdots (p-2) \equiv 1 \pmod{p}.$$

Hence

$$(p-1)! \equiv 1 \cdot 2 \cdot 3 \cdots (p-2)(p-1) \equiv 1 \cdot 1 \cdot (p-1) \equiv -1 \pmod{p},$$

and we have proved half of the theorem. It remains to prove the other half and show that if

(3) $$(n-1)! \equiv -1 \pmod{n},$$

then n is a prime. Suppose that $n = ab$ for some integers a and b, with $a \ne n$. From (3), we have

$$n \mid (n-1)! + 1,$$

and since $a \mid n$, we have

(4) $$a \mid (n-1)! + 1.$$

But since $a \le n - 1$, it follows that one of the factors of $(n-1)!$ is a itself. Thus

(5) $$a \mid (n - 1)!.$$

But (4) and (5) imply that $a \mid 1$. Hence the only positive divisors of n are 1 and n, and thus n is a prime.

Problems

* 1. What is the least residue of

$$5^6 \text{ (mod 7)} \qquad 5^8 \text{ (mod 7)} \qquad 1945^8 \text{ (mod 7)?}$$

2. What is the least residue of

$$5^{10} \text{ (mod 11)} \qquad 5^{12} \text{ (mod 11)} \qquad 1945^{12} \text{ (mod 11)?}$$

* 3. What is the last digit of 7^{355}?
4. What are the last two digits of 7^{355}?
* 5. What is the remainder when 314^{162} is divided by 163?
6. What is the remainder when 314^{162} is divided by 7?
* 7. What is the remainder when 314^{164} is divided by 165? (Watch out—165 is not prime!)
8. What is the remainder when 2001^{2001} is divided by 26?
9. Show that

$$(p - 1)(p - 2) \cdots (p - r) \equiv (-1)^r r! \text{(mod } p),$$

for $r = 1, 2, \ldots, p - 1$.

10. (a) Calculate $(n - 1)! \pmod{n}$ for $n = 10, 12, 14,$ and 15.
 (b) Guess a theorem and prove it.
† 11. Show that $2(p - 3)! + 1 \equiv 0 \pmod{p}$.

12. In 1732 Euler wrote: "I derived [certain] results from the elegant theorem, of whose truth I am certain, although I have no proof: $a^n - b^n$ is divisible by the prime $n + 1$ if neither a nor b is." Prove this theorem, using Fermat's Theorem.

13. Note that

$$6! \equiv -1 \text{ (mod 7),}$$
$$5!1! \equiv 1 \text{ (mod 7),}$$
$$4!2! \equiv -1 \text{ (mod 7),}$$
$$3!3! \equiv 1 \text{ (mod 7).}$$

Try the same sort of calculation (mod 11).

14. Guess a theorem from the data of Problem 13, and prove it.
† 15. Suppose that p is an odd prime.
 (a) Show that

$$1^{p-1} + 2^{p-1} + \cdots + (p-1)^{p-1} \equiv -1 \pmod{p}.$$

(b) Show that

$$1^p + 2^p + \cdots + (p-1)^p \equiv 0 \pmod{p}.$$

16. Show that the converse of Fermat's Theorem is false. [Broad hint: consider $2^{340} \pmod{341}$.]

† 17. Show that for any two different primes p, q,
 (a) $pq \mid (a^{p+q} - a^{p+1} - a^{q+1} + a^2)$ for all a.
 (b) $pq \mid (a^{pq} - a^p - a^q + a)$ for all a.

18. Show that if p is an odd prime, then $2p \mid (2^{2p-1} - 2)$.

*† 19. For what n is it true that

$$p \mid (1 + n + n^2 + \cdots + n^{p-2})?$$

20. Show that every odd prime except 5 divides some number of the form 111 . . . 11 (k digits, all ones).

7

The Divisors of an Integer

It would be natural now to continue studying congruences by taking up quadratic congruences, but partly for the sake of variety we will take up a different subject and return to congruences later.

Let n be a positive integer. Let $d(n)$ denote the **number of positive divisors of** n (including 1 and n), and let $\sigma(n)$, denote **the sum of the positive divisors of** n. That is,

$$d(n) = \sum_{d \mid n} 1 \quad \text{and} \quad \sigma(n) = \sum_{d \mid n} d.$$

$\sum_{d \mid n}$ means **the sum over the positive divisors of** n. For example,

$$\sum_{d \mid 12} h(d) = h(1) + h(2) + h(3) + h(4) + h(6) + h(12)$$

and

$$\sum_{d \mid 10} d^2 = 1 + 4 + 25 + 100.$$

These functions occur frequently, and in this section we will derive some of their properties. In the next section we will use them to study perfect, abundant, and other kinds of numbers first considered by the ancient Greek mathematicians and still of interest today.

* **Exercise 1.** Verify that the following table is correct as far as it goes, and complete it.

n	1	2	3	4	5	6	7	8	9	10	11	12	13	14	15	16
$d(n)$	1	2	2	3	2	4	2	4	3	4						

If p is a prime, then $d(p) = 2$, because the only positive divisors of p are 1 and p. Since p^2 has divisors $1, p,$ and p^2, then $d(p^2) = 3$.

* **Exercise 2.** What is $d(p^3)$? Generalize to $d(p^n)$, $n = 4, 5, \ldots$.

If $p \neq q$, then pq has divisors $1, p, q,$ and pq, so $d(pq) = 4$. (In this section, as elsewhere, p and q will stand for primes.) Similarly, the divisors of p^2q are $1, p, p^2, q, pq,$ and p^2q, so $d(p^2q) = 6$.

* **Exercise 3.** What is $d(p^3q)$? What is $d(p^nq)$ for any positive n?

After Exercises 2 and 3, you may have guessed

Theorem 1. If $p_1^{e_1}p_2^{e_2} \cdots p_k^{e_k}$ is the prime-power decomposition of n, then

$$d(n) = d(p_1^{e_1})d(p_2^{e_2}) \cdots d(p_k^{e_k}).$$

Proof. Let D denote the set of numbers

(1) $$p_1^{f_1}p_2^{f_2} \cdots p_k^{f_k}, \qquad 0 \leq f_i \leq e_i.$$

We claim that D is exactly the set of divisors of n. First, we note that every number in the set is a divisor of n, because we can find for each number in (1) an integer whose product with the number is n; namely,

$$p_1^{e_1-f_1}p_2^{e_2-f_2} \cdots p_k^{e_k-f_k}.$$

Second, suppose that d is a divisor of n. If $p \mid d$, then $p \mid n$, so each prime in the prime-power decomposition of d must appear in the prime-power decomposition of n. Thus

$$d = p_1^{f_1}p_2^{f_2} \cdots p_k^{f_k},$$

where some (or all) of the exponents may be 0. Moreover, no exponent f_i is larger than e_i. (If it were, we would have a situation in which $p_i^{f_i} \mid d$ and $d \mid n$, which implies $p_i^{f_i} \mid n$. This is impossible if $f_i > e_i$.) Thus every divisor of n is a member of the set D. Thus D is identical with the set of divisors of n. Each f_i in (1) may take on $e_i + 1$ values. Thus there are

$$(e_1 + 1)(e_2 + 1)(e_3 + 1) \cdots (e_k + 1)$$

numbers in D, and because of the unique factorization theorem, they are all different. Since $d(p^n) = n + 1$, this proves the theorem. For example, from $24 = 2^3 \cdot 3$ we get $d(24) = d(2^3)d(3) = 4 \cdot 2 = 8$.

* **Exercise 4.** Calculate $d(240)$.

Now we will get a formula for $\sigma(n)$.

* **Exercise 5.** Verify that the following table is correct as far as it goes, and complete it.

n	1	2	3	4	5	6	7	8	9	10	11	12	13	14
$\sigma(n)$	1	3	4	7	6	12	8	15						

As with $d(n)$, some special cases are easy. For example, $\sigma(p) = p + 1$ for all primes p. Furthermore, the divisors of p^2 are $1, p,$ and p^2, so $\sigma(p^2) = 1 + p + p^2$.

* **Exercise 6.** What is $\sigma(p^3)$? $\sigma(pq)$, where p and q are different primes?

Exercise 7. Show that $\sigma(2^n) = 2^{n+1} - 1$.

* **Exercise 8.** What is $\sigma(p^n)$, $n = 1, 2, \ldots$?

Let us calculate $\sigma(p^e q^f)$, where p and q are different primes, and see if it suggests a general result. The divisors of $p^e q^f$ are

$$
\begin{array}{ccccc}
1 & p & p^2 & \cdots & p^e, \\
q & pq & p^2 q & \cdots & p^e q, \\
q^2 & pq^2 & p^2 q^2 & \cdots & p^e q^2, \\
 & & \cdot & & \\
 & & \cdot & & \\
 & & \cdot & & \\
q^f & pq^f & p^2 q^f & \cdots & p^e q^f.
\end{array}
$$

If we add across each row, we get

$$
\begin{aligned}
\sigma(p^e q^f) &= (1 + p + \cdots + p^e) + q(1 + p + \cdots + p^e) \\
&\quad + q^2(1 + p + \cdots + p^e) + \cdots \\
&\quad + q^f(1 + p + \cdots + p^e) \\
&= (1 + q + \cdots + q^f)(1 + p + \cdots + p^e) = \sigma(q^f)\sigma(p^e).
\end{aligned}
$$

What is true for the product of two prime powers is true in general:

Theorem 2. If $p_1^{e_1} p_2^{e_2} \cdots p_k^{e_k}$ is the prime-power decomposition of n, then

$$\sigma(n) = \sigma(p_1^{e_1})\sigma(p_2^{e_2}) \cdots \sigma(p_k^{e_k}).$$

Proof. We will use mathematical induction. The theorem is true for $k = 1$ and, as we have just seen, for $k = 2$. Suppose that it is true for $k = r$. We will show that this implies that it is true for $k = r + 1$. Let

$$n = p_1^{e_1} p_2^{e_2} \cdots p_r^{e_r} p_{r+1}^{e_{r+1}} = N p_{r+1}^{e_{r+1}}.$$

To simplify the notation, let us write $n = Np^e$. Let $1, d_1, \ldots, d_t$ be the divisors of N. Since $(N, p) = 1$, all of the divisors of n are

$$
\begin{array}{ccccc}
1 & d_1 & d_2 & \cdots & d_t \\
p & d_1 p & d_2 p & \cdots & d_t p \\
 & & \cdot & & \\
 & & \cdot & & \\
 & & \cdot & & \\
p^e & d_1 p^e & d_2 p^e & \cdots & d_t p^e
\end{array}
$$

Summing, we get

$$\sigma(n) = (1 + d_1 + d_2 + \cdots + d_t)(1 + p + \cdots + p^e)$$
$$= \sigma(N)\sigma(p^e).$$

But from the induction assumption,

$$\sigma(N) = \sigma(p_1^{e_1})\sigma(p_2^{e_2}) \cdots \sigma(p_k^{e_k}),$$

and the last two equations complete the proof.

For example, from $24 = 2^3 \cdot 3$ we get

$$\sigma(24) = \sigma(2^3)\sigma(3) = (1 + 2 + 4 + 8)(1 + 3) = 60.$$

* **Exercise 9.** Calculate $\sigma(240)$.

Both d and σ are members of an important class of number-theoretic functions: the multiplicative functions. We will now define this term, verify that d and σ are multiplicative functions, and explain why the idea is important. A function f, defined for the positive integers, is said to be **multiplicative** if and only if

$$(m, n) = 1 \quad \text{implies} \quad f(mn) = f(m)f(n).$$

A simple example of a multiplicative function is given by $f(n) = n$.
Another is g, where $g(n)$ is the product of the prime divisors of n.

Theorem 3. d is multiplicative.

Proof. Let m and n be relatively prime. Then, no prime that divides
m can divide n, and vice versa. Thus if

$$m = p_1^{e_1}p_2^{e_2} \cdots p_k^{e_k} \qquad \text{and} \qquad n = q_1^{f_1}q_2^{f_2} \cdots q_r^{f_r}$$

are the prime-power decompositions of m and n, then no q is a p
and no p is a q, and the prime-power decomposition of mn is given by

$$mn = p_1^{e_1}p_2^{e_2} \cdots p_k^{e_k}q_1^{f_1}q_2^{f_2} \cdots q_r^{f_r}.$$

Applying Theorem 1, we have

$$d(mn) = d(p_1^{e_1})d(p_2^{e_2}) \cdots d(p_k^{e_k})d(q_1^{f_1})d(q_2^{f_2}) \cdots d(q_r^{f_r})$$
$$= d(p_1^{e_1}p_2^{e_2} \cdots p_k^{e_k})d(q_1^{f_1}q_2^{f_2} \cdots q_r^{f_r}) = d(m)d(n).$$

Theorem 4. σ is multiplicative.

Proof. The proof is exactly the same as the proof of Theorem 3,
with d replaced by σ and "Theorem 1" by "Theorem 2."

The reason multiplicative functions are important is that if we know
the value of a multiplicative function f for all prime-powers, then we
can find the value of f for all positive integers. To see this, we note

Theorem 5. If f is a multiplicative function and the prime-power
decomposition of n is $p_1^{e_1}p_2^{e_2} \cdots p_k^{e_k}$, then

$$f(n) = f(p_1^{e_1})f(p_2^{e_2}) \cdots f(p_k^{e_k}).$$

Proof. The proof is by induction on k. The theorem is trivially true
for $k = 1$. Suppose it is true for $k = r$. Because

$$(p_1^{e_1}p_2^{e_2} \cdots p_r^{e_r}, p_{r+1}^{e_{r+1}}) = 1,$$

we have, from the definition of a multiplicative function,

$$f((p_1^{e_1}p_2^{e_2} \cdots p_r^{e_r})p_{r+1}^{e_{r+1}}) = f(p_1^{e_1}p_2^{e_2} \cdots p_r^{e_r})f(p_{r+1}^{e_{r+1}}).$$

From the induction assumption, the first factor is

$$f(p_1^{e_1}p_2^{e_2} \cdots p_r^{e_r}) = f(p_1^{e_1})f(p_2^{e_2}) \cdots f(p_r^{e_r}),$$

and this, together with the preceding equation, completes the induction.

For an example, suppose that $f(p^e) = ep^{e-1}$ for all primes p and all e, $e \geq 1$. The first few values of f are

n	2	3	4	5	6	7	8	9	10	11	12
$f(n)$	1	1	4	1	1	1	12	6	1	1	4

$$f(3141) = f(3^2 \cdot 349) = f(3^2)f(349) = 6 \cdot 1 = 6,$$

and we can calculate $f(n)$ for any n in a similar manner.

* **Exercise 10.** Compute $f(n)$ for $n = 13, 14, \ldots, 24$.

We will apply Theorem 5 of this section in Section 9 to get a formula for an important number-theoretic function, Euler's ϕ-function.

Problems

* 1. Calculate $d(42)$, $\sigma(42)$, $d(420)$, and $\sigma(420)$.

 2. Calculate $d(540)$, $\sigma(540)$, $d(5400)$, and $\sigma(5400)$.

* 3. Calculate d and σ of $10115 = 5 \cdot 7 \cdot 17^2$ and $100115 = 5 \cdot 20023$.

 4. Calculate d and σ of $10116 = 2^2 \cdot 3^2 \cdot 281$ and $100116 = 2^2 \cdot 3^5 \cdot 103$.

 5. Show that $\sigma(n)$ is odd if n is a power of two.

 6. Prove that if $f(n)$ is multiplicative, then so is $f(n)/n$.

* 7. What is the smallest integer n such that $d(n) = 8$? Such that $d(n) = 10$?

 8. Does $d(n) = k$ have a solution n for each k?

* 9. In 1644, Mersenne asked for a number with 60 divisors. Find one smaller than 10,000.

 10. Find infinitely many n such that $d(n) = 60$.

* 11. If p is an odd prime, for which k is $1 + p + \cdots + p^k$ odd?

 12. For which n is $\sigma(n)$ odd?

 13. If n is a square, show that $d(n)$ is odd..

 14. If $d(n)$ is odd, show that n is a square.

† 15. Observe that $1 + 1/3 = 4/3$; $1 + 1/2 + 1/4 = 7/4$; $1 + 1/5 = 6/5$; $1 + 1/2 +$

1/3 + 1/6 = 15/6; 1 + 1/7 = 8/7; and 1 + 1/2 + 1/4 + 1/8 = 15/8. Guess and prove a theorem.

16. Find infinitely many n such that $\sigma(n) \le \sigma(n-1)$.

*† 17. If N is odd, how many solutions does $x^2 - y^2 = N$ have?

18. Develop a formula for $\sigma_2(n)$, the sum of the squares of the positive divisors of n.

† 19. Guess a formula for

$$\sigma_k(n) = \sum_{d \mid n} d^k,$$

where k is a positive integer.

20. Show that the product of the positive divisors of n is $n^{d(n)/2}$.

Section

Perfect Numbers

Numbers are fascinating. At least, they have always fascinated some people, from the time of the start of mathematics as a deductive art some 2600 years ago until today. The ancient Greek mathematicians mixed number theory with mysticism, and numbers could be male or female, square or triangular, abundant or deficient, and they took on nonnumerical properties. For example, 5 was the number of marriage because it was the sum of the smallest male and female numbers greater than 1: $5 = 2 + 3$. Number mysticism has never disappeared, and though today it takes forms different from the past, it is still with us, flourishing for the same reasons as astrology.

A number is called **perfect** if and only if it is equal to the sum of its positive divisors, excluding itself. For example, 6 is perfect, because $6 = 1 + 2 + 3$. So is 28 perfect, because $28 = 1 + 2 + 4 + 7 + 14$. But 18 is not perfect, because the sum of its positive divisors, excluding itself, is $1 + 2 + 3 + 6 + 9 = 21$. We study perfect numbers not for mystical reasons, but because they provide practice with the σ-function and because Euler proved a satisfying theorem that allows us to determine all even perfect numbers. Long before Euler, Euclid found some perfect numbers; we will follow in his footsteps.

In symbols, the sum of the positive divisors of n, excluding itself, is

$\sigma(n) - n$. Hence a number is perfect if and only if $\sigma(n) = 2n$. To find solutions to this equation, we will need to use a result proved in Section 7—namely, that σ is a multiplicative function, or

$$(1) \qquad \text{if} \quad (m, n) = 1, \qquad \text{then} \qquad \sigma(mn) = \sigma(m)\sigma(n).$$

With its aid, we can prove

Theorem 1 *(Euclid)*. If $2^k - 1$ is prime, then $2^{k-1}(2^k - 1)$ is perfect.

Proof. Let $n = 2^{k-1}(2^k - 1)$. Because $2^k - 1$ is prime, we know that $\sigma(2^k - 1) = 2^k$. Then, noting that 2^{k-1} and $2^k - 1$ are relatively prime and applying (1), we have

$$\sigma(n) = \sigma(2^{k-1}(2^k - 1)) = \sigma(2^{k-1})\sigma(2^k - 1) = (2^k - 1) \cdot 2^k = 2n.$$

Thus n is perfect.

So, every time we find a k such that $2^k - 1$ is prime, we can construct a perfect number. We do not have to look at all values of k, because if k is composite, so is $2^k - 1$: if $k = ab$, then

$$2^k - 1 = 2^{ab} - 1 = (2^a - 1)(2^{a(b-1)} + 2^{a(b-2)} + \cdots + 1).$$

Thus $2^k - 1$ can be prime only when k is prime. Replacing k with p, the first few such numbers are

p	2	3	5	7	11	13
$2^p - 1$	3	7	31	127	2047	8191

and all of these except $2047 = 23 \cdot 89$ are prime. Thus we have five perfect numbers:

$$2(2^2 - 1) = 6,$$
$$2^2(2^3 - 1) = 28,$$
$$2^4(2^5 - 1) = 496,$$
$$2^6(2^7 - 1) = 8128,$$
$$2^{12}(2^{13} - 1) = 33550336.$$

An example of a larger perfect number is

191561942608236107294793378084303638130993721548169216.

Now we will show that the numbers $2^{p-1}(2^p - 1)$ with p and $2^p - 1$ prime are the only even perfect numbers.

Theorem 2 (Euler). **If** n **is an even perfect number, then**

$$n = 2^{p-1}(2^p - 1)$$

for some prime p, and $2^p - 1$ is also prime.

Proof. If n is an even perfect number, then $n = 2^e m$, where m is odd and $e \geq 1$. Since $\sigma(m) > m$, we can write $\sigma(m) = m + s$, with $s > 0$. Then $2n = \sigma(n)$ becomes

$$2^{e+1} m = (2^{e+1} - 1)(m + s) = 2^{e+1}m - m + (2^{e+1} - 1)s.$$

Thus

$$(2) \qquad\qquad m = (2^{e+1} - 1)s,$$

which says that s is a divisor of m and $s < m$. But $\sigma(m) = m + s$; thus s is the *sum* of all the divisors of m that are less than m. That is, s is the sum of a group of numbers that includes s. This is possible only if the group consists of one number alone. Therefore, the set of divisors of m smaller than m must contain only one element, and that element must be 1. That is, $s = 1$, and hence $m = 2^{e+1} - 1$ is a prime.

We repeat the argument, because it is slippery. Let the divisors of m be

$$1, d_2, d_3, \ldots, d_k, m.$$

Then $\sigma(m) = m + s$, or

$$s = 1 + d_2 + d_3 + \cdots + d_k.$$

But s is a divisor of m and $s < m$, so s *equals* one of $1, d_2, \ldots, d_k$. The only way that can be possible is if $s = 1$.

We have shown that $s = 1$. Thus $\sigma(m) = m + s = m + 1$. This says that m is prime. From (2), $m = 2^{e+1} - 1$. The only numbers of this form that can be prime are those with $e + 1$ prime. Hence $m = 2^p - 1$ for some prime p, and this completes the proof.

Thus the even perfect numbers determined in Theorem 1 are the only even perfect numbers. As for odd perfect numbers, no one knows if there are any, and no one has proved that none can exist. It is known that if there is an odd perfect number, then it is quite large: it must be greater than 10^{38}, and it must satisfy many other conditions. For example, if p_1, p_2, \ldots, p_k are the prime factors of an odd perfect number, then $1/p_1 + 1/p_2 + \cdots + 1/p_k > (150/151)\ln 2$. But no combination of

conditions has so far served to show that there are no odd perfect numbers: it may be that there is one, but so huge that it is out of the range of human computation.

The problem of finding even perfect numbers is, after Theorem 2, the same as the problem of determining primes p such that $2^p - 1$ is also prime. Primes of the form $2^p - 1$ are called *Mersenne primes*. In the seventeenth century, Mersenne, a lover of mathematics, claimed that $2^p - 1$ was prime for

$$k = 2, 3, 5, 7, 13, 17, 31, 67, 127, 257,$$

and for no other primes less than 257. His guess was not accurate: he erred in including 67 and 257, for

$$2^{67} - 1 = 193707721 \cdot 761838257287,$$

and $2^{257} - 1$ is also composite. He further erred in excluding 19, 61, 89, and 107. But let us not think harshly of Mersenne: in the seventeenth century, there were no mathematical journals to announce new discoveries; instead, almost everyone wrote to Mersenne, and Mersenne wrote to almost everyone else, enclosing the latest mathematical news. He thus spread the results of Fermat, for one, and sped the development of mathematics; it is fitting that he have a set of primes named after him. The complete list of currently known primes p such that $2^p - 1$ is prime is

$$2, 3, 5, 7, 13, 17, 19, 31, 61, 89, 107, 127, 521, 607, 1279,$$
$$2203, 2281, 3217, 4253, 4423, 9689, 9941, 11213, 19937,$$

and to each of these there corresponds an even perfect number: 24 in all. The first twelve were discovered before the invention of high-speed computers; the later ones are so enormous as to be beyond the reach of hand computation. The search for Mersenne primes has gone on (the last one was announced in 1971), in the hope of seeing some sort of pattern in the primes p, so that theorems could be guessed and maybe proved. Some conjectures have been advanced, but without any indication as to how to go about proving them, and no important theorems have been proved. It is not even known if there are infinitely many such primes.

To close the section, we will mention another kind of number that arose from number mysticism: *amicable numbers*. Consider 220 and 284. Since $220 = 2^5 \cdot 5 \cdot 11$, it follows that

$$\sigma(220) - 220 = \sigma(2^2)\sigma(5)\sigma(11) - 220 = 7 \cdot 6 \cdot 12 - 220$$
$$= 504 - 220 = 284.$$

And since $284 = 2^2 \cdot 71$, we have

$$\sigma(284) - 284 = \sigma(2^2)\sigma(71) - 284 = 7 \cdot 72 - 284$$
$$= 504 - 284 = 220.$$

So, in some sense, 220 and 284 go together. In general, we say that m and n are **amicable** (or are an **amicable pair**) if and only if

$$\sigma(m) - m = n \qquad \text{and} \qquad \sigma(n) - n = m.$$

Equivalently, we could say that m and n are amicable if and only if

$$\sigma(m) = \sigma(n) = m + n.$$

Exercise 1. Verify that 1184 and 1210 are amicable.

Number mystics think that if one person carries a talisman of some sort containing the number 220, and another person has one with 284, they will be favorably disposed to each other. Numbers undeniably have power: it might be worth a try. The amicable pair in Exercise 1 was first discovered as late as 1866, Euler found many such pairs, and long lists of them exist. Besides those already mentioned, the amicable pairs less than 10,000 are 2620, 2924; 5020, 5564; and 6232, 6368. There are as yet no general theorems on amicable numbers as beautiful as Euclid's and Euler's theorems on perfect numbers. Perhaps they remain to be discovered.

In the problems, abundant, deficient, and triangular numbers appear. An integer n is **abundant** if and only if $\sigma(n) - n > n$, and it is **deficient** if and only if $\sigma(n) - n < n$. **Triangular** numbers have the form $n(n + 1)/2$.

Problems

1. Verify that 2620, 2924 and 17296, 18416 are amicable pairs. (The latter pair, discovered by Fermat, was the second pair found. Note that $17296 = 2^4 \cdot 23 \cdot 47$ and $18416 = 2^4 \cdot 1151$.)

2. It was long thought that even perfect numbers ended alternately in 6 and 8. Show that this is wrong by verifying that the perfect numbers corresponding to the primes $2^{13} - 1$ and $2^{17} - 1$ both end in 6.

* 3. Classify the integers 2, 3, . . . , 21 as abundant, deficient, or perfect.

4. Classify the integers 402, 403, . . . , 421 as abundant, deficient, or perfect.

5. If $\sigma(n) = kn$, then n is called a *k-perfect number*. Verify that 672 is 3-perfect and $2{,}178{,}540 = 2^2 \cdot 3^2 \cdot 5 \cdot 7^2 \cdot 13 \cdot 19$ is 4-perfect.

6. Show that no number of the form $2^a 3^b$ is 3-perfect.

7. Let us say that n is *superperfect* if and only if $\sigma(\sigma(n)) = 2n$. Show that if $n = 2^k$ and $2^{k+1} - 1$ is prime, then n is superperfect.

8. It was long thought that every abundant number was even. Show that 945 is abundant, and find another abundant number of the form $3^a \cdot 5 \cdot 7$.

9. In 1575, it was observed that every even perfect number is a triangular number. Show that this is so.

10. In 1652, it was observed that

$$6 = 1 + 2 + 3,$$
$$28 = 1 + 2 + 3 + 4 + 5 + 6 + 7,$$
$$496 = 1 + 2 + 3 + \cdots + 31.$$

Can this go on?

11. Let

$$p = 3 \cdot 2^e - 1,$$
$$q = 3 \cdot 2^{e-1} - 1,$$
$$r = 3^2 \cdot 2^{2e-1} - 1,$$

where e is a positive integer. If p, q, and r are all prime, show that $2^e pq$ and $2^e r$ are amicable. (Only for $e = 2, 4, 7$ are p, q, and r all prime for $e \le 200$.)

12. Show that if $p > 3$ and $2p + 1$ is prime, then $2p(2p + 1)$ is deficient.

13. Show that all even perfect numbers end in 6 or 8.

14. If n is an even perfect number and $n > 6$, show that the sum of its digits is congruent to 1 (mod 9).

15. If p is odd, show that $2^{p-1}(2^p - 1) \equiv 1 + 9p(p - 1)/2 \pmod{81}$.

9

Euler's Theorem and Function

Fermat's Theorem states that if p is prime, then

$$(a, p) = 1 \quad \text{implies} \quad a^{p-1} \equiv 1 \ (\text{mod } p).$$

It is natural to ask if there is a generalization of this to any positive integer: Given any integer m, is there a number $f(m)$ such that $a^{f(m)} \equiv 1$ (mod m)? We note that this cannot hold unless $(a, m) = 1$, for if a and m have a common divisor greater than 1, then $m \mid (a^k - 1)$ is impossible for any $k > 0$. Let us look at tables of powers of a (mod m), where a and m are relatively prime, for $m = 6$, 9, and 10:

$m = 6$			$m = 9$						$m = 10$				
a	a^2		a	a^2	a^3	a^4	a^5	a^6		a	a^2	a^3	a^4
1	1		1	1	1	1	1	1		1	1	1	1
5	1		2	4	8	7	5	1		3	9	7	1
			4	7	1	4	7	1		7	9	3	1
			5	7	8	4	2	1		9	1	9	1
			7	4	1	7	4	1					
			8	1	8	1	8	1					

Evidently,

$$a^2 \equiv 1 \text{ (mod 6)} \qquad \text{if} \qquad (a, 6) = 1,$$
$$a^6 \equiv 1 \text{ (mod 9)} \qquad \text{if} \qquad (a, 9) = 1,$$
$$a^4 \equiv 1 \text{ (mod 10)} \qquad \text{if} \qquad (a, 10) = 1,$$

so the number $f(m)$ exists for $m = 6, 9$, and 10.

Exercise 1. Show that $a^6 \equiv 1$ (mod 14) for all a relatively prime to 14.

If your eye is very sharp indeed, you might have noticed that $f(6) = 2$ and that there are two positive integers less than 6 and relatively prime to 6; $f(9) = 6$, and there are six positive integers less than 9 and relatively prime to it; $f(10) = 4$, and there are four positive integers less than 10 and relatively prime to it; and a similar statement holds for 14.

Let us introduce some notation. If m is a positive integer, let $\phi(m)$ denote the number of positive integers less than or equal to m and relatively prime to m. We will call ϕ **Euler's ϕ-function.** With this notation, $\phi(6) = 2$, $\phi(9) = 6$, and $\phi(10) = 4$. If you worked more examples you would almost certainly guess that the following theorem, first proved by Euler, is true:

Theorem 1. Suppose that $m \geq 1$ and $(a, m) = 1$. Then $a^{\phi(m)} \equiv 1$ (mod m).

We know that the theorem is true in the special case when $m = p$, a prime. Every positive integer less than p is relatively prime to it, so $\phi(p) = p - 1$, and it is Fermat's Theorem that $a^{p-1} \equiv 1$ (mod p) when $(a, p) = 1$. The theorem is also true if $m = 1$, since the definition of the ϕ-function gives $\phi(1) = 1$.

In this section we will prove Theorem 1, and we will develop a formula for calculating $\phi(n)$ from the prime-power decomposition of n.

The idea used to prove Fermat's Theorem is that if $(a, p) = 1$, then the least residues (mod p) of $a, 2a, \ldots, (p - 1)a$ are a permutation of $1, 2, \ldots, p - 1$. This is also the key to Euler's generalization:

Lemma 1. If $(a, m) = 1$ and $r_1, r_2, \ldots, r_{\phi(m)}$ are the positive integers less than m and relatively prime to m, then the least residues (mod m) of

(1) $$ar_1, ar_2, \ldots, ar_{\phi(m)}$$

are a permutation of

$$r_1, r_2, \ldots, r_{\phi(m)}.$$

For example, if $m = 10$ and $a = 3$, then the least residues of

$$3 \cdot 1, \quad 3 \cdot 3, \quad 3 \cdot 7, \quad 3 \cdot 9$$

are

$$3, \quad 9, \quad 1, \quad 7,$$

a permutation of $\{1, 3, 7, 9\}$.

Exercise 2. Verify that Lemma 1 is true if $m = 14$ and $a = 5$.

Proof of Lemma 1. Since there are exactly $\phi(m)$ numbers in the set (1), to prove that their least residues are a permutation of the $\phi(m)$ numbers $r_1, r_2, \ldots, r_{\phi(m)}$ we have to show that they are all different and that they are all relatively prime to m. To show that they are all different, suppose that

$$ar_i \equiv ar_j \pmod{m}$$

for some i and j ($1 \le i \le \phi(m)$, $1 \le j \le \phi(m)$). Since $(a, m) = 1$, we can cancel a from both sides of the congruence to get $r_i \equiv r_j \pmod{m}$. Since r_i and r_j are least residues (mod m), it follows that $r_i = r_j$. Hence, $r_i \ne r_j$ implies $ar_i \ne ar_j \pmod{m}$, and so the numbers in (1) are all different.

To prove that all the numbers in (1) are relatively prime to m, suppose that p is a prime common divisor of ar_i and m for some i, $1 \le i \le \phi(m)$. Since p is prime, either $p \,|\, a$ or $p \,|\, r_i$. Thus either p is a common divisor of a and m or of r_i and m. But $(a, m) = (r_i, m) = 1$, so both cases are impossible. Hence $(ar_i, m) = 1$ for each i, $i = 1, 2, \ldots, \phi(m)$.

The proof of Euler's Theorem proceeds as does the proof of Fermat's Theorem:

Proof of Theorem 1. From Lemma 1 we know that

$$r_1 r_2 \cdots r_{\phi(m)} \equiv (ar_1)(ar_2) \cdots (ar_{\phi(m)})$$
$$\equiv a^{\phi(m)}(r_1 r_2 \cdots r_{\phi(m)}) \pmod{m}.$$

Since each of $r_1, r_2, \ldots, r_{\phi(m)}$ is relatively prime to m, it follows that their product is, also; thus that factor may be canceled in the last congruence, and we get

$$1 \equiv a^{\phi(m)} \pmod{m}.$$

The rest of the section will be mainly devoted to the properties of ϕ; our goal is to find a way of calculating $\phi(n)$ by some method other than actually counting all the positive integers less than n and relatively prime to it.

Exercise 3. Verify that the entries in the following table are correct.

n	2	3	4	5	6	7	8	9	10
$\phi(n)$	1	2	2	4	2	6	4	6	4

Exercise 4. Verify that $3^{\phi(8)} \equiv 1 \pmod{8}$.

* **Exercise 5.** Which positive integers are less than 4 and relatively prime to it? What is the answer if 4 is replaced by 8? By 16? Can you induce a formula for $\phi(2^n)$, $n = 1, 2, \ldots$?

In general, it is not hard to see what $\phi(p^n)$ is, where p is a prime and n is a positive integer.

Lemma 2. $\phi(p^n) = p^{n-1}(p - 1)$ for all positive integers n.

Proof. The positive integers less than or equal to p^n which are *not* relatively prime to p^n are exactly the multiples of p:

$$1 \cdot p, 2 \cdot p, 3 \cdot p, \ldots, (p^{n-1})p,$$

and there are p^{n-1} of them. Since there are in all p^n positive integers less than or equal to p^n, we have

$$\phi(p^n) = p^n - p^{n-1} = p^{n-1}(p - 1).$$

For example, the positive integers less than or equal to 27 which are not relatively prime to 27 are 3, 6, 9, 12, 15, 18, 21, 24, and 27; nine of them, so $\phi(27) = 27 - 9 = 9(3 - 1)$.

Exercise 6. Verify that the formula is correct for $p = 5$ and $n = 2$.

Thus we know ϕ for all prime-powers. If we knew that ϕ was a multiplicative function, then we could apply Theorem 5 of Section 7 to get a formula for $\phi(n)$. That ϕ is in fact multiplicative we will now demonstrate in a theorem whose proof, in common with many other proofs in number theory, is neither long, technical, nor complicated; it is just hard. First we need an easy lemma:

Lemma 3. If $(a, m) = 1$ and $a \equiv b \pmod{m}$, then $(b, m) = 1$.

Proof. This follows from the fact that $b = a + km$ from some k.

Corollary. If the least residues (mod m) of

(2) $$r_1, r_2, \ldots, r_m$$

are a permutation of $0, 1, \ldots, m - 1$, then (2) contains exactly $\phi(m)$ elements relatively prime to m.

We can now prove

Theorem 2. ϕ is multiplicative.

Proof. Suppose that $(m, n) = 1$ and write the numbers from 1 to mn as follows:

$$
\begin{array}{ccccc}
1 & m+1 & 2m+1 & \ldots & (n-1)m+1 \\
2 & m+2 & 2m+2 & \ldots & (n-1)m+2 \\
 & & \vdots & & \\
m & 2m & 3m & \ldots & mn
\end{array}
$$

Suppose that $(m, r) = d$ and $d > 1$. Then we claim that no element in the rth row of the array:

$$r \quad m+r \quad 2m+r \quad \ldots \quad km+r \quad \ldots \quad (n-1)m+r$$

is relatively prime to mn. This is so because if $d \mid m$ and $d \mid r$, then $d \mid (km - r)$ for any k. So, if we are looking for numbers that are rela-

tively prime to mn, we will not find any except in those rows whose first element is relatively prime to m.

* **Exercise 7.** How many such rows are there?

For an example, let us take $n = 5$ and $m = 6$. Then the array is

$$
\begin{array}{ccccc}
1 & 7 & 13 & 19 & 25 \\
2 & 8 & 14 & 20 & 26 \\
3 & 9 & 15 & 21 & 27 \\
4 & 10 & 16 & 22 & 28 \\
5 & 11 & 17 & 23 & 29 \\
6 & 12 & 18 & 24 & 30
\end{array}
$$

No element in the second, third, fourth, or sixth row is relatively prime to $mn = 30$, because the first element in each of those rows is not relatively prime to $m = 6$. All numbers relatively prime to 30 are found in the two remaining rows,

(3)
$$
\begin{array}{ccccc}
1 & 7 & 13 & 19 & 25 \\
5 & 11 & 17 & 23 & 29
\end{array}
$$

Suppose we can show that there are exactly $\phi(n)$ numbers relatively prime to mn in each of the rows that have first elements relatively prime to m. Since there are $\phi(m)$ such rows, it will follow that the number of integers in the whole array that are relatively prime to mn is $\phi(n)\phi(m)$: that is, $\phi(mn) = \phi(m)\phi(n)$, and the theorem will be proved. But the numbers in the rth row (where r and m are relatively prime) are

(4) $r, m + r, 2m + r, \ldots, (n - 1)m + r,$

and we claim that their least residues (mod n) are a permutation of

(5) $0, 1, 2, \ldots, (n - 1).$

To verify this claim, all we have to do is show that no two of the numbers in (4) are congruent (mod n), because (4) contains n elements, just as (5) does. This is easy: suppose that

$$km + r \equiv jm + r \pmod{n},$$

with $0 \le k < n$ and $0 \le j < n$. Then $km \equiv jm \pmod{n}$, and since $(m, n) = 1$, we have $k \equiv j \pmod{n}$. On account of the inequalities on k and j,

it follows that $k = j$. Hence, if $k \neq j$, then $km + r \not\equiv jm + r \pmod{n}$, and no two elements of (4) are congruent \pmod{n}.

In the example above, the least residues $\pmod 5$ of the numbers in (3) are

$$\begin{array}{ccccc} 1 & 2 & 3 & 4 & 0 \\ 0 & 1 & 2 & 3 & 4 \end{array}$$

and each row contains $\phi(5) = 4$ numbers relatively prime to 30. Thus $8 = \phi(30) = \phi(6)\phi(5)$.

By the Corollary to Lemma 3, we have that (4) contains exactly $\phi(n)$ elements relatively prime to n. But from Lemma 3, every element in the rth row of the array is relatively prime to m. It follows that the rth row of the array contains exactly $\phi(n)$ elements relatively prime to mn. As we noted before, this is enough to complete the proof.

We can now get a formula for $\phi(n)$:

Theorem 3. If n has a prime-power decomposition given by

$$n = p_1^{e_1} p_2^{e_2} \cdots p_k^{e_k},$$

then

$$\phi(n) = p_1^{e_1-1}(p_1 - 1)p_2^{e_2-1}(p_2 - 1) \cdots p_k^{e_k-1}(p_k - 1).$$

Proof. Because ϕ is multiplicative, Theorem 5 of Section 7 applies here to give

$$\phi(n) = \phi(p_1^{e_1})\phi(p_2^{e_2}) \cdots \phi(p_k^{e_k}).$$

If we apply Lemma 2 to each term on the right, the theorem is proved.

The best way to calculate a value for ϕ is to use this multiplicative property. For example, to find $\phi(72)$, factor 72 into $2^3 \cdot 3^2$ and write $\phi(72) = \phi(2^3)\phi(3^2) = 2^2 \cdot 1 \cdot 3^1 \cdot 2 = 24$.

* **Exercise 8.** Calculate $\phi(74)$, $\phi(76)$, and $\phi(78)$.

The formula of Theorem 3 can be written in another form, which is neater and sometimes useful, though not for computation:

Corollary. If $n = p_1^{e_1} p_2^{e_2} \cdots p_k^{e_k}$, then

$$\phi(n) = n \left(1 - \frac{1}{p_1} \right) \left(1 - \frac{1}{p_2} \right) \cdots \left(1 - \frac{1}{p_k} \right).$$

The proof of this corollary is left to the reader. We conclude with a theorem we will need in the next section.

* **Exercise 9.** Calculate $\sum_{d \mid n} \phi(d)$

 (a) For $n = 12, 13, 14, 15$, and 16.
 (b) For $n = 2^k, k \geq 1$.
 (c) For $n = p^k, k \geq 1$ and p an odd prime.

You should have guessed by now that the following theorem is true.

Theorem 4. If $n \geq 1$, then

$$\sum_{d \mid n} \phi(d) = n.$$

Proof. It would be natural to try to apply the formula of Theorem 3 to get this result. This would be difficult; instead we use a clever idea first thought of by Gauss. Consider the integers $1, 2, \ldots, n$. We will put one of these integers in class C_d if and only if its greatest common divisor with n is d. For example, if $n = 12$, we have

$$C_1 = \{1, 5, 7, 11\}, \qquad C_2 = \{2, 10\},$$
$$C_3 = \{3, 9\}, \qquad C_4 = \{4, 8\},$$
$$C_6 = \{6\}, \qquad C_{12} = \{12\}.$$

* **Exercise 10.** What are the classes C_d for $n = 14$?

We have m in C_d if and only if $(m, n) = d$. But $(m, n) = d$ if and only if $(m/d, n/d) = 1$. That is, an integer m is in class C_d if and only if m/d is relatively prime to n/d. The number of positive integers less than or equal to n/d and relatively prime to n/d is $\phi(n/d)$ by definition. Thus, the number of elements in class C_d is $\phi(n/d)$.

Exercise 11. Check that this is correct for $n = 12$ and $n = 14$.

Since there is a class for each divisor of n, the total number of elements in all the classes C_d is

$$\sum_{d|n} \phi(n/d).$$

That is, $n = \sum_{d|n} \phi(n/d)$. But $\sum_{d|n} \phi(n/d)$ is the same as $\sum_{d|n} \phi(n)$. For example, the positive divisors of 9 are 1, 3, and 9, so $\sum_{d|9} \phi(d) = \phi(1) + \phi(3) + \phi(9)$ and $\sum_{d|9} \phi(n/d) = \phi(9) + \phi(3) + \phi(1)$; the same terms in the opposite order. Hence $n = \sum_{d|n} \phi(d)$, and the theorem is proved.

Problems

* 1. Calculate $\phi(42)$, $\phi(420)$, and $\phi(4200)$.

2. Calculate $\phi(54)$, $\phi(540)$, and $\phi(5400)$.

* 3. Calculate ϕ of $10115 = 5 \cdot 7 \cdot 17^2$ and $100115 = 5 \cdot 20023$.

4. Calculate ϕ of $10116 = 2^2 \cdot 3^2 \cdot 281$ and $100116 = 2^2 \cdot 3^5 \cdot 103$.

* 5. Calculate $a^8 \pmod{15}$ for $a = 1, 2, \ldots, 14$.

6. Calculate $a^8 \pmod{16}$ for $a = 1, 2, \ldots, 15$.

7. Show that if n is odd, then $\phi(4n) = 2\,\phi(n)$.

8. Perfect numbers satisfy $\sigma(n) = 2n$. Which n satisfy $\phi(n) = 2n$?

† 9. $1 + 2 = (3/2)\phi(3)$, $1 + 3 = (4/2)\phi(4)$, $1 + 2 + 3 + 4 = (5/2)\phi(5)$, $1 + 5 = (6/2)\phi(6)$, $1 + 2 + 3 + 4 + 5 + 6 = (7/2)\phi(7)$, and $1 + 3 + 5 + 7 = (8/2)\phi(8)$. Guess a theorem.

10. Show that

$$\sum_{p \le x} \sigma(p) - \sum_{p \le x} \phi(p) = \sum_{p \le x} d(p).$$

11. Prove Lemma 3 by starting with the fact that there are integers r and s such that $ar + ms = 1$.

12. If $(a, m) = 1$, show that any x such that

$$x \equiv ca^{\phi(m)-1} \pmod{m}$$

satisfies $ax \equiv c \pmod{m}$.

13. Let $f(n) = (n + \phi(n))/2$. Show that $f(f(n)) = \phi(n)$ if $n = 2^k, k = 2, 3, \ldots$

14. Find four solutions of $\phi(n) = 16$.

*† 15. Find all solutions of $\phi(n) = 4$ and prove that there are no more.

16. Show that $\phi(mn) > \phi(m)\phi(n)$ if m and n have a common factor greater than 1.

† 17. Show that $(m, n) = 2$ implies $\phi(mn) = 2\,\phi(m)\phi(n)$.

18. Show that $\phi(n) = n/2$ if and only if $n = 2^k$ for some positive integer k.

† 19. Show that if $n - 1$ and $n + 1$ are both primes and $n > 4$, then $\phi(n) \leq n/3$.

20. Show that $\phi(n) = 14$ is impossible.

10

Primitive Roots

In Theorem 1 of the last section, we saw that if $(a, m) = 1$, then there is a positive integer t such that $a^t \equiv 1 \pmod{m}$, namely $t = \phi(m)$. This can be proved independently of that theorem as follows. If $(a, m) = 1$, then the least residues (mod m) of a, a^2, a^3, . . . are all relatively prime to m. There are $\phi(m)$ least residues (mod m) that are relatively prime to m and infinitely many powers of a: it follows that there are positive integers j and k such that $j \neq k$ and $a^j \equiv a^k \pmod{m}$. Since $(a, m) = 1$, the smaller power of a in the last congruence may be canceled, and we have either

$$a^{j-k} \equiv 1 \pmod{m} \qquad \text{or} \qquad a^{k-j} \equiv 1 \pmod{m}.$$

So, if $(a, m) = 1$, then there is a positive integer t such that $a^t \equiv 1 \pmod{m}$. In fact there are infinitely many, since it follows from $a^{\phi(m)} \equiv 1 \pmod{m}$ for $(a, m) = 1$ that for any positive integer k,

$$a^{t+k\phi(m)} \equiv a^t (a^k)^{\phi(m)} \equiv a^t \equiv 1 \pmod{m}.$$

The smallest such t is important enough that it has a name: if $(a, m) = 1$, then the **order** of a modulo m is the smallest positive integer t such that $a^t \equiv 1 \pmod{m}$.

$$a^n \ (\text{mod } 11)$$

a	a^2	a^3	a^4	a^5	a^6	a^7	a^8	a^9	a^{10}
1	1	1	1	1	1	1	1	1	1
2	4	8	5	10	9	7	3	6	1
3	9	5	4	1	3	9	5	4	1
4	5	9	3	1	4	5	9	3	1
5	3	4	9	1	5	3	4	9	1
6	3	7	9	10	5	8	4	2	1
7	5	2	3	10	4	6	9	8	1
8	9	6	4	10	3	2	5	7	1
9	4	3	5	1	9	4	3	5	1
10	1	10	1	10	1	10	1	10	1

For example, from the table above,

2, 6, 7, and 8 have order 10 (mod 11),

3, 4, 5, and 9 have order 5,

10 has order 2, and

1 has order 1.

When this idea was introduced by Gauss, he called t *the exponent to which a belongs* (mod m), and his long phrase is sometimes used in place of our shorter one.

* **Exercise 1.** What are the orders of 3, 5, and 7, modulo 8?

Note that in the above example, the orders 1, 2, 5, and 10 are divisors of $\phi(11) = 10$. This is no coincidence, as we show in Theorem 2. First we show that the set of integers n for which $a^n \equiv 1 \ (\text{mod } m)$ is precisely the set of the multiples of the order of a (mod m).

Theorem 1. Suppose that $(a, m) = 1$ and a has order t (mod m). Then $a^n \equiv 1 \ (\text{mod } m)$ if and only if n is a multiple of t.

Proof. Suppose that $n = tq$ for some integer q. Then $a^n \equiv a^{tq} \equiv (a^t)^q \equiv 1^q \equiv 1 \ (\text{mod } m)$ because $a^t \equiv 1 \ (\text{mod } m)$.

Conversely, suppose that $a^n \equiv 1 \ (\text{mod } m)$. Since t is the smallest positive integer such that $a^t \equiv 1 \ (\text{mod } m)$, we have $n \geq t$ so we can divide n by t to get $n = tq + r$ with $q \geq 1$ and $0 \leq r < t$. Thus

$$1 \equiv a^n \equiv a^{tq+r} \equiv (a^t)^q a^r \equiv 1^q a^r \equiv a^r \ (\text{mod } m).$$

Since t is the smallest positive integer such that $a^t \equiv 1 \pmod{m}$, $a^r \equiv 1$ \pmod{m} with $0 \le r < t$ is possible only if $r = 0$. Thus $n = tq$ and the theorem is proved.

This gives

Theorem 2. If $(a, m) = 1$ and a has order $t \pmod{m}$, then $t \mid \phi(m)$.

Proof. From Euler's extension of Fermat's Theorem we know that $a^{\phi(m)} \equiv 1 \pmod{m}$. From Theorem 1, $\phi(m)$ is a multiple of t, which is what we wanted to prove.

* **Exercise 2.** What order can an integer have (mod 9)? Find an example of each.

As an example of an application of this idea, we prove

Theorem 3. If p and q are odd primes and $q \mid a^p - 1$, then either $q \mid a - 1$ or $q = 2kp + 1$ for some integer k.

Proof. Since $q \mid a^p - 1$, we have $a^p \equiv 1 \pmod{q}$. So, by Theorem 1, the order of $a \pmod{q}$ is a divisor of p. That is, a has order 1 or p. If the order of a is 1, then $a^1 \equiv 1 \pmod{q}$, so $q \mid a - 1$. If on the other hand the order of a is p, then by Theorem 2, $p \mid \phi(q)$; that is, $p \mid q - 1$. So, $q - 1 = rp$ for some integer r. Since p and q are odd, r must be even, and this completes the proof.

Corollary. Any divisor of $2^p - 1$ is of the form $2kp + 1$.

* **Exercise 3.** Using the corollary, what is the smallest possible prime divisor of $2^{19} - 1$?

The next theorem is essentially a corollary to Theorem 1, but it is worth stating explicitly.

Theorem 4. If the order of $a \pmod{m}$ is t, then $a^r \equiv a^s \pmod{m}$ if and only if $r \equiv s \pmod{t}$.

Proof. Suppose that $a^r \equiv a^s$ (mod m). We can suppose that $r \geq s$ with no loss of generality. Thus $a^{r-s} \equiv 1$ (mod m), and from Theorem 1, $r - s$ is a multiple of t. That, by definition, says that $r \equiv s$ (mod t).

To prove the converse, suppose that $r \equiv s$ (mod t). Then $r = s + kt$ for some integer k, and

$$a^r \equiv a^{s+kt} \equiv a^s(a^t)^k \equiv a^s \text{ (mod } m)$$

because $a^t \equiv 1$ (mod m).

If a is a least residue and the order of a (mod m) is $\phi(m)$, then we will say that a is a **primitive root** of m. Primitive roots, and numbers that have them, are of special interest because of the following property:

Theorem 5. If g is a primitive root of m, then the least residues, modulo m, of

$$g, g^2, \ldots, g^{\phi(m)}$$

are a permutation of the $\phi(m)$ positive integers less than m and relatively prime to it.

Proof. Since $(g, m) = 1$, each power of g is relatively prime to m. Moreover, no two powers have the same least residue, because if $g^j \equiv g^k$ (mod m), then from Theorem 4, $j \equiv k$ (mod $\phi(m)$). If $j \not\equiv k$ (mod $\phi(m)$), $g^j \not\equiv g^k$ (mod m).

For example, 2 is a primitive root of 9, and the powers

$$2, 2^2, 2^3, 2^4, 2^5, 2^6$$

are, (mod 9),

$$2, 4, 8, 7, 5, 1.$$

Exercise 4. Show that 3 is a primitive root of 7.

* **Exercise 5.** Find, by trial, a primitive root of 10.

Not every integer has primitive roots—for example, 8 does not, as we saw in Exercise 1. We will now set out to show that each prime has a primitive root. The proof is not easy, requires a good deal of prepara-

tion (Lemmas 1 through 3), and because it is an existence proof, it does not show how to find the primitive root. For these reasons, you do not lose too much if you take the result on faith.

If a has order t (mod m), then any power of a will have an order no larger than t, because for any k, $(a^k)^t \equiv (a^t)^k \equiv 1$ (mod m). From the table of a^n (mod 11) we can see that 2 is a primitive root of 11, and among its powers, 2^3, 2^7, and 2^9 have order 10; 2^2, 2^4, 2^6, and 2^8 have order 5; and 2^5 has order 2. Lemma 1 tells which powers of a have the same order as a.

Lemma 1. Suppose that a has order t (mod m). Then a^k has order t (mod m) if and only if $(k, t) = 1$.

Proof. Suppose that $(k, t) = 1$, and denote the order of a^k by s. We have

$$1 \equiv (a^t)^k \equiv (a^k)^t \pmod{m},$$

so from Theorem 1, $s \mid t$. Because s is the order of a^k,

$$(a^k)^s \equiv a^{ks} \equiv 1 \pmod{m},$$

so from Theorem 1 again, $t \mid ks$. Since $(k, t) = 1$, it follows that $t \mid s$. This fact, together with the fact that $s \mid t$, implies $s = t$.

To prove the converse, suppose that a and a^k have order t and that $(k, t) = r$. Then

$$1 \equiv a^t \equiv (a^t)^{k/r} \equiv (a^k)^{t/r} \pmod{m};$$

because t is the order of a^k, Theorem 1 says that t/r is a multiple of t. This implies that $r = 1$.

For example, 2 has order 10 (mod 11), and the lemma says that 2^k has order 10 if and only if $(k, 10) = 1$; that is, for $k = 1, 3, 7,$ and 9. The other primitive roots of 11 are thus 2^3, 2^7, and 2^9, or 8, 7, and 6. Thus, if we can find one primitive root of a prime, we can find them all. This is worth stating as a

Corollary. Suppose that g is a primitive root of p. Then the least residue of g^k is a primitive root of p if and only if $(k, p - 1) = 1$.

Proof. Apply Lemma 1 with $t = p - 1$.

We now need a lemma about the solutions of polynomial congruences (mod p). Though it is an important theorem, we do not use it elsewhere.

Lemma 2. If f is a polynomial of degree n, then

(1) $f(x) \equiv 0 \pmod{p}$

has at most n solutions.

Proof. Let

$$f(x) = a_n x^n + a_{n-1} x^{n-1} + \cdots + a_0$$

have degree n; that is, $a_n \not\equiv 0 \pmod{p}$. We prove the lemma by induction. For $n = 1$,

$$a_1 x + a_0 \equiv 0 \pmod{p}$$

has but one solution, since $(a_1, p) = 1$. Suppose that the lemma is true for polynomials of degree $n - 1$, and suppose that f has degree n. Either $f(x) \equiv 0 \pmod{p}$ has no solutions or it has at least one. In the first case, the lemma is true. In the second case, suppose that r is a solution. That is, $f(r) \equiv 0 \pmod{p}$, and r is a least residue (mod p). Then because $x - r$ is a factor of $x^t - r^t$ for $t = 0, 1, \ldots, n$, we have

$$\begin{aligned} f(x) &\equiv f(x) - f(r) \\ &\equiv a_n(x^n - r^n) + a_{n-1}(x^{n-1} - r^{n-1}) + \cdots + a_1(x - r) \\ &\equiv (x - r)g(x) \pmod{p}, \end{aligned}$$

where g is a polynomial of degree $n - 1$. Suppose that s is also a solution of (1). Thus

$$f(s) \equiv (s - r)g(s) \equiv 0 \pmod{p}.$$

Because p is a prime it follows that

$$s \equiv r \pmod{p} \quad \text{or} \quad g(s) \equiv 0 \pmod{p};$$

from the induction assumption, the second congruence has at most $n - 1$ solutions. Since the first congruence has just one solution, the proof is complete.

Note that Lemma 2 is not true if the modulus is not a prime. For example,

$$x^2 + x \equiv 0 \pmod{6}$$

has solutions 0, 2, 3, and 5.

Lemma 3. If $d \mid p - 1$, then $x^d \equiv 1 \pmod{p}$ has exactly d solutions.

Proof. From Fermat's Theorem, the congruence $x^{p-1} \equiv 1 \pmod{p}$ has exactly $p - 1$ solutions, namely $1, 2, \ldots, p - 1$. Moreover,

$$x^{p-1} - 1 = (x^d - 1)(x^{p-1-d} + x^{p-1-2d} + \cdots + 1)$$
$$= (x^d - 1)h(x).$$

From Lemma 2, we know that $h(x) \equiv 0 \pmod{p}$ has at most $p - 1 - d$ solutions. Hence $x^d \equiv 1 \pmod{p}$ has at least d solutions. Applying Lemma 2 again, we see that it has exactly d solutions.

We are at last prepared to prove

Theorem 6. Every prime p has $\phi(p - 1)$ primitive roots.

Proof. Theorem 2 says that each of the integers

$$(2) \qquad\qquad\qquad 1, 2, \ldots, p - 1$$

has an order that is a divisor of $p - 1$. For each divisor t of $p - 1$, let $\psi(t)$ denote the number of integers in (2) that have order t. Restating what we have just said:

$$\sum_{t \mid p-1} \psi(t) = p - 1.$$

From Theorem 4 of Section 9, we have

$$(3) \qquad\qquad\qquad \sum_{t \mid p-1} \psi(t) = \sum_{t \mid p-1} \phi(t).$$

If we can show that $\psi(t) \le \phi(t)$ for each t, it will follow from (3) that $\psi(t) = \phi(t)$ for each t. In particular, the number of primitive roots of p will be $\psi(p - 1) = \phi(p - 1)$.

Choose some t. If $\psi(t) = 0$, then $\psi(t) < \phi(t)$ and we are done. If $\psi(t) \ne 0$, then there is an integer with order t; call it a. The congruence

$$(4) \qquad\qquad\qquad x^t \equiv 1 \pmod{p}$$

has, according to Lemma 3, exactly t solutions. Furthermore, (4) is satisfied by the t integers

$$(5) \qquad\qquad\qquad a, a^2, a^3, \ldots, a^t,$$

and because no two of these have the same least residue \pmod{p}, they give all the solutions. From Lemma 1, the numbers in (5) that have order t are those powers a^k with $(k, t) = 1$. But there are $\phi(t)$ such numbers k. Hence $\psi(t) = \phi(t)$ in this case. As noted above, this completes the proof.

We have actually proved more than was stated in Theorem 6. Although we will not use what we have proved, we have the

Corollary. If p is a prime and $t \mid (p - 1)$, then the number of least residues (mod p) with order t is $\phi(t)$.

Exercise 6. Use the table of powers (mod 11) at the beginning of this section to verify that the corollary is true for $p = 11$.

Theorem 6 does not actually help us to find a primitive root of a prime. To find one, we may use tables or trial. Here is a table giving the smallest positive primitive root, g_p, for each prime p less than 114.

p	2	3	5	7	11	13	17	19	23	29	31	37	41	43	47
g_p	1	2	2	3	2	2	3	2	5	2	3	2	6	3	5

p	53	59	61	67	71	73	79	83	89	97	101	103	107	109	113
g_p	2	2	2	2	7	5	3	2	3	5	2	5	2	6	3

No method is known for predicting what will be the smallest positive primitive root of a given prime p, nor is there much known about the distribution of the $\phi(p - 1)$ primitive roots among the least residues modulo p. For example, the primitive roots of 71 and 73 are

Primitive roots of 71						Primitive roots of 73					
7	11	13	21	22	28	5	11	13	14	15	20
31	33	35	42	44	47	26	28	29	31	33	34
52	53	55	56	59	61	39	40	42	44	45	47
62	63	65	67	68	69	53	58	59	60	62	68

There are other numbers besides primes that have primitive roots. It can be proved that the only positive integers with primitive roots are 1, 2, 4, p^e, and $2p^e$, where p is an odd prime and e is a positive integer.

Exercise 7. Which of the integers 2, 3, . . . , 25 do not have primitive roots?

As an example of the application of primitive roots, we will use them to prove part of Wilson's Theorem quickly and elegantly. Let g be a primitive root of the odd prime p. From Theorem 5, we know that the least residues (mod p) of g, g^2, \ldots, g^{p-1} are a permutation of 1, 2, . . . , $p - 1$. Multiplying and using the fact that

$$1 + 2 + 3 + \cdots + (p - 1) = (p - 1)p/2,$$

we get

$$1 \cdot 2 \cdots (p - 1) \equiv g \cdot g^2 \cdots g^{p-1} \pmod{p}$$

or

$$(p - 1)! \equiv (g^p)^{(p-1)/2} \equiv g^{(p-1)/2} \pmod{p}.$$

But $g^{(p-1)/2}$ satisfies $x^2 \equiv 1 \pmod{p}$, and we know that $g^{(p-1)/2} \equiv 1$ or -1 (mod p). But the first case is impossible, since g is a primitive root of p. Thus $(p - 1)! \equiv -1 \pmod{p}$.

Problems

* 1. Find the orders of 1, 2, . . . , 12 (mod 13).

2. Find the orders of 1, 2, . . . , 16 (mod 17).

* 3. One of the primitive roots of 19 is 2. Find all of the others.

4. One of the primitive roots of 23 is 5. Find all of the others.

* 5. What are the orders of 2, 4, 7, 8, 11, 13, and 14 (mod 15)? Does 15 have primitive roots?

6. What are the orders of 3, 7, 9, 11, 13, 17, and 19 (mod 20)? Does 20 have primitive roots?

* 7. Which integers have order 6 (mod 31)?

8. Which integers have order 6 (mod 37)?

9. If a, $a \neq 1$, has order t (mod p), show that

$$a^{t-1} + a^{t-2} + \cdots + 1 \equiv 0 \pmod{p}.$$

10. If g and h are primitive roots of an odd prime p, then $g \equiv h^k \pmod{p}$ for some integer k. Show that k is odd.

† 11. Show that if g and h are primitive roots of an odd prime p, then the least residue of gh is not a primitive root of p.

12. If g, h, and k are primitive roots of p, is the least residue of ghk always a primitive root of p?

† 13. Show that if a has order 3 (mod p), then $a + 1$ has order 6 (mod p).

14. If p and q are odd primes and $q \mid a^p + 1$, show that either $q \mid a + 1$ or $q = 2kp + 1$ for some integer k.

*† 15. Suppose that a has order 4 (mod p). What is the least residue of $(a + 1)^4$ (mod p)?

16. Show that $131071 = 2^{17} - 1$ is prime.

† 17. Show that $(2^{19} + 1)/3$ is prime.

18. If g is a primitive root of p, show that two consecutive powers of g have

consecutive least residues. That is, show that there exists k such that $g^{k+1} \equiv g^k + 1 \pmod{p}$.

† 19. If g is a primitive root of p, show that no three consecutive powers of g have consecutive least residues. That is, show that $g^{k+2} \equiv g^{k+1} + 1 \equiv g^k + 2 \pmod{p}$ is impossible for any k.

20. (a) Show that if m is a number having primitive roots, then the product of the positive integers less than or equal to m and relatively prime to it is congruent to $-1 \pmod{m}$.

(b) Show that the result in (a) is not always true if m does not have primitive roots.

Quadratic Congruences

After studying linear congruences, it is natural to look at quadratic congruences:

$$Ax^2 + Bx + C \equiv 0 \ (\text{mod } m).$$

In this section we will restrict the modulus to an odd prime. We will assume that $A \not\equiv 0 \ (\text{mod } p)$, because if $A \equiv 0 \ (\text{mod } p)$, then

(1) $$Ax^2 + Bx + C \equiv 0 \ (\text{mod } p)$$

would be a linear congruence, not a quadratic congruence. We know that there is an integer A' such that $AA' \equiv 1 \ (\text{mod } p)$. Hence (1) has the same solutions as

(2) $$x^2 + A'Bx + A'C \equiv 0 \ (\text{mod } p).$$

* **Exercise 1.** Convert $2x^2 + 3x + 1 \equiv 0 \ (\text{mod } 5)$ to a quadratic congruence whose first coefficient is 1.

If $A'B$ is even, we can complete the square in (2) to get

$$\left(x + \frac{A'B}{2} \right)^2 \equiv \left(\frac{A'B}{2} \right)^2 - A'C \ (\text{mod } p);$$

if $A'B$ is odd, we can change it to $p + A'B$, which is even, and then

83

complete the square. In either case, we have replaced (1) with an equivalent congruence of the form

(3) $y^2 \equiv a \pmod{p}$;

thus, if we can solve this congruence, we can solve any quadratic congruence \pmod{p}. For example, to solve $3x^2 + 6x + 5 \equiv 0 \pmod{7}$, first multiply both sides of the congruence by 5 and reduce the coefficients $\pmod{7}$ to get $x^2 + 2x + 4 \equiv 0 \pmod{7}$. Then complete the square: $x^2 + 2x + 1 \equiv 4 \pmod{7}$ or $(x + 1)^2 \equiv 4 \pmod{7}$. Since $y^2 \equiv 4$ $\pmod{7}$ has the solutions 2 and 5, the original quadratic congruence has solutions 1 and 4.

* **Exercise 2.** Change the quadratic in Exercise 1 to the form (3).

* **Exercise 3** (optional). By inspection, find all the solutions of the congruence in Exercise 2.

Such congruences do not always have solutions. For example, modulo 5,

$$0^2 \equiv 0, \qquad 1^2 \equiv 4^2 \equiv 1, \qquad \text{and} \qquad 2^2 \equiv 3^2 \equiv 4,$$

so $x^2 \equiv a \pmod{5}$ has a solution for $a = 0$, 1, or 4 and no solution for $a = 2$ or 3. We note that $x^2 \equiv 0 \pmod{p}$ has only the solution $x \equiv 0 \pmod{p}$. We now show that if $p \nmid a$, then solutions of $x^2 \equiv a \pmod{p}$ come in pairs. This should be no surprise: since $r^2 = (-r)^2$, we have $r^2 \equiv (-r)^2$ \pmod{p}, so if r is a solution of $x^2 \equiv a \pmod{p}$, then so is the least residue \pmod{p} of $-r$. Thus if r is a solution, so is $p - r$.

Theorem 1. Suppose that p is an odd prime. If $p \nmid a$, then $x^2 \equiv a \pmod{p}$ has exactly two solutions or no solutions.

Proof. Suppose that the congruence has a solution, and call it r. Then $p - r$ is a solution too, and it is different from r. (For if $r \equiv p - r \pmod{p}$, then $2r \equiv 0 \pmod{p}$; since $(2, p) = 1$, we get $r \equiv 0 \pmod{p}$, which is impossible.) Let s be any solution. Then $r^2 \equiv s^2 \pmod{p}$, whence $p \mid (r - s)(r + s)$. Thus

$$p \mid (r - s) \qquad \text{or} \qquad p \mid (r + s).$$

In the first case, $s \equiv r \pmod{p}$. In the second case, $s \equiv p - r \pmod{p}$. Since s, r, and $p - r$ are all least residues, we have $s = r$ or $p - r$; these are thus the only solutions.

This theorem is not true if the modulus is not prime. For example, $x^2 \equiv 1 \pmod 8$ has four solutions, namely 1, 3, 5, and 7.

* **Exercise 4.** If $p > 3$, what are the two solutions of $x^2 \equiv 4 \pmod p$?

It follows from Theorem 1 that if a is selected from the integers 1, 2, . . . , $p - 1$, then $x^2 \equiv a \pmod p$ will have two solutions for $(p - 1)/2$ values of a and no solutions for the other $(p - 1)/2$ values of a. For example, if $p = 11$, then x^2 is one of the entries in the table

x	1	2	3	4	5	6	7	8	9	10
x^2 (mod 11)	1	4	9	5	3	3	5	9	4	1

since $x^2 \equiv (p - x)^2 \pmod p$, the entries are symmetric about $p/2$ and the same $(p - 1)/2$ least residues appear in each half. So, for those $(p - 1)/2$ least residues, $x^2 \equiv a \pmod p$ has two solutions, and for the other $(p - 1)/2$ least residues there are no solutions.

* **Exercise 5.** For what values of a does $x^2 \equiv a \pmod 7$ have two solutions?

It would be nice to be able to tell the two groups apart. In this section we will do this by deriving **Euler's Criterion:**

Theorem 2. If p is an odd prime and $p \nmid a$, then $x^2 \equiv a \pmod p$ has a solution or no solution depending on whether

$$a^{(p-1)/2} \equiv 1 \qquad \text{or} \qquad -1 \pmod p.$$

First we introduce some new words.

If $x^2 \equiv a \pmod m$ has a solution, then a is called a **quadratic residue** (mod m).

If $x^2 \equiv a \pmod m$ has no solution, then a is called a **quadratic non-residue** (mod m).

There are also cubic residues, fourth-power residues, and so on, but assuming there is no danger of confusion, we will omit the adjective "quadratic" and refer to residues and nonresidues for short.

Euler's Criterion can be easily derived using primitive roots.

Proof of Theorem 2. Let g be a primitive root of the odd prime p. Then $a \equiv g^k \pmod p$ for some k. If k is even, then $x^2 \equiv a \pmod p$ has a

solution, namely the least residue of $g^{k/2}$; further, by Fermat's Theorem,

$$a^{(p-1)/2} \equiv (g^k)^{(p-1)/2} \equiv (g^{k/2})^{(p-1)} \equiv 1 \ (\text{mod } p).$$

If k is odd, then

$$a^{(p-1)/2} \equiv (g^{(p-1)/2})^k \equiv (-1)^k \equiv -1 \ (\text{mod } p),$$

and also $x^2 \equiv a \ (\text{mod } p)$ has no solution: if it did have one, say r, we would have

$$1 \equiv r^{p-1} \equiv (r^2)^{(p-1)/2} \equiv a^{(p-1)/2} \equiv -1 \ (\text{mod } p),$$

which is impossible.

As an example of the application of the criterion, let us see if $x^2 \equiv 7$ (mod 31) has a solution. We must calculate $7^{(31-1)/2} = 7^{15}$ and see what its remainder is upon division by 31. Of course we do not need to carry out the actual division: we have

$$7^2 \equiv 49 \equiv 18 \ (\text{mod } 31);$$

squaring, we get

$$7^4 \equiv 18^2 \equiv 324 \equiv 14 \ (\text{mod } 31),$$
$$7^8 \equiv 14^2 \equiv 196 \equiv 10 \ (\text{mod } 31),$$

and

$$7^{16} \equiv 10^2 \equiv 100 \equiv 7 \ (\text{mod } 31).$$

Since 7 and 31 are relatively prime, we may divide the last congruence by 7 to get $7^{15} \equiv 1 \ (\text{mod } 31)$. It follows from Euler's Criterion that $x^2 \equiv 7 \ (\text{mod } 31)$ has a solution.

Though Euler's Criterion tells us when $x^2 \equiv a \ (\text{mod } p)$ has solutions, it gives us no way of actually finding them. Of course, it is possible to substitute $x = 1, 2, 3, 4, \ldots$ until a solution is found, but this procedure can be long and tiresome. The following method—adding multiples of the modulus and factoring squares—is sometimes more convenient. For example, take $x^2 \equiv 7 \ (\text{mod } 31)$, which we know to have a solution. Adding 31 repeatedly, we have

$$x^2 \equiv 7 \equiv 38 \equiv 69 \equiv 100 \equiv 10^2 \ (\text{mod } 31),$$

and we see immediately that the congruence is satisfied when $x = 10$ or -10; the two solutions are thus 10 and 21. That example was easy; a

more typical one is $x^2 \equiv 41 \pmod{61}$, which Euler's Criterion shows to have a solution. We have

$$x^2 \equiv 41 \equiv 102 \equiv 163 \equiv 224 \equiv 4^2 \cdot 14 \pmod{61}.$$

Also,

$$14 \equiv 75 \equiv 5^2 \cdot 3 \pmod{61},$$

so $x^2 \equiv 4^2 \cdot 5^2 \cdot 3 \pmod{61}$. But

$$3 \equiv 64 \equiv 8^2 \pmod{61},$$

so $x^2 \equiv 4^2 \cdot 5^2 \cdot 8^2 \equiv 160^2 \equiv 38^2 \pmod{61}$. Thus $x \equiv \pm 38 \pmod{61}$, and the two solutions are 38 and 23. This method will, with more or less labor, always produce the solutions.

* **Exercise 6.** Find the solutions of $x^2 \equiv 8 \pmod{31}$.

Euler's Criterion is sometimes cumbersome to apply, even to congruences with small numbers like $x^2 \equiv 3201 \pmod{8191}$. We will now develop a method for deciding when an integer is a quadratic residue \pmod{p}. The method is relatively easy to apply, even when the numbers are 3201 and 8191. It is based on the famous quadratic reciprocity theorem, which has many applications other than the one we will use it for.

We start by introducing notation to abbreviate the long phrase, "$x^2 \equiv a \pmod{p}$ has a solution." The French mathematician A. M. Legendre thought that he had proved the quadratic reciprocity theorem (Theorem 4 of this section). He was mistaken, but in the course of his work on it, he introduced a useful symbol:

The **Legendre symbol,** (a/p), where p is an odd prime and $p \nmid a$ is defined by

$$(a/p) = \begin{cases} 1 & \text{if } a \text{ is a quadratic residue } \pmod{p} \\ -1 & \text{if } a \text{ is a quadratic nonresidue } \pmod{p}. \end{cases}$$

For example, $(3/5) = -1$ because $x^2 \equiv 3 \pmod{5}$ has no solutions, and $(1/5) = 1$, because 1 is a quadratic residue $\pmod{5}$. Neither $(7/15)$ nor $(91/7)$ is defined, the first because the second entry in the symbol is not an odd prime, and the second because $7 \mid 91$.

* **Exercise 7.** What is $(1/3)$? $(1/7)$? $(1/11)$? In general, what is $(1/p)$?

* **Exercise 8.** What is $(4/5)$? $(4/7)$? $(4/p)$ for any odd prime p?

* **Exercise 9** (optional). Induce a theorem from the two preceding exercises.

To find out whether $x^2 \equiv 3201 \pmod{8191}$ has a solution, we can evaluate $(3201/8191)$. To do this, we will need some rules on how Legendre symbols can be manipulated. We will start with three simple but important properties.

Theorem 3. The Legendre symbol has the properties

 (A) if $a \equiv b \pmod{p}$, then $(a/p) = (b/p)$,

 (B) if $p \nmid a$, then $(a^2/p) = 1$,

 (C) if $p \nmid a$ and $p \nmid b$, then $(ab/p) = (a/p)(b/p)$.

In the above properties and throughout the rest of this section, we will agree that p and q represent odd primes, and that the first entry in a Legendre symbol is not a multiple of the second entry; with these conventions, all Legendre symbols are defined.

Proof of Theorem 3. (A): Suppose that $x^2 \equiv a \pmod{p}$ has a solution. If $a \equiv b \pmod{p}$, then $x^2 \equiv b \pmod{p}$ also has a solution—the same one. This shows that

(4) if $(a/p) = 1$ and $a \equiv b \pmod{p}$, then $(b/p) = 1$.

Exercise 10. Verify that

(5) if $(a/p) = -1$ and $a \equiv b \pmod{p}$, then $(b/p) = -1$.

Together, (4) and (5) show that (A) is true.

 (B): Clearly, $x^2 \equiv a^2 \pmod{p}$ has a solution—namely, the least residue of $a \pmod{p}$.

 (C): This important property of the Legendre symbol, in combination with the quadratic reciprocity theorem, makes the symbol useful for computations. In words, (C) says that the product of two residues is a residue; the product of two nonresidues is a residue; and the product of a residue and a nonresidue is a nonresidue. To prove (C) we use Euler's Criterion. In terms of the Legendre symbol, it says

$$(a/p) = 1 \quad \text{if} \quad a^{(p-1)/2} \equiv 1 \pmod{p},$$

and

$$(a/p) = -1 \quad \text{if} \quad a^{(p-1)/2} \equiv -1 \pmod{p}.$$

Comparing the 1's and −1's, we see that

(6) $$(a/p) \equiv a^{(p-1)/2} \pmod{p}.$$

So, from (6) and the fact that $(xy)^n \equiv x^n y^n \pmod{p}$, we have

$$(ab/p) \equiv (ab)^{(p-1)/2} \equiv a^{(p-1)/2} b^{(p-1)/2} \equiv (a/p)(b/p) \pmod{p}.$$

We have not yet proved (C); we have only shown that

$$(ab/p) \equiv (a/p)(b/p) \pmod{p}.$$

But the left-hand side of the congruence is either 1 or −1, and so is the right-hand side. Hence, the only way that the two numbers can be congruent modulo p is if they are equal. We have now proved (C).

We can also use (6) to give quick proofs of (A) and (B). For example, to prove (A), we have from (6)

$$(a/p) \equiv a^{(p-1)/2} \equiv b^{(p-1)/2} \equiv (b/p) \pmod{p},$$

and since the value of a Legendre symbol is 1 or −1, congruence implies equality.

Exercise 11. Prove (B), using (6).

Exercise 12. Prove that $(4a/p) = (a/p)$.

* **Exercise 13.** Evaluate $(19/5)$ and $(-9/13)$ by using (A) and (B).

The quadratic reciprocity theorem tells us how (p/q) and (q/p) are related. The theorem was guessed by Euler and Legendre years before it was first proved by Gauss, who eventually gave several proofs. It is an example of a deep and important theorem whose statement was arrived at by observation. Consider the following tables:

p

q	5	7	11	13	17	19	23
3	−1	1	−1	1	−1	1	−1
5		−1	1	−1	−1	1	−1
7			1	−1	−1	−1	1
11				−1	−1	−1	−1
13					1	−1	1
17						1	−1
19							1

(p/q)

p

q	5	7	11	13	17	19	23
3	−1	−1	1	1	−1	−1	1
5		−1	1	−1	−1	1	−1
7			−1	−1	−1	1	−1
11				−1	−1	1	1
13					1	−1	1
17						1	−1
19							−1

(q/p)

Can you by observation see any relation between (p/q) and (q/p)? These tables are perhaps too small to allow any firm guesses to be made (Gauss's tables had thousands of entries), but note that the columns in both tables are the same for $p = 5$, 13, and 17. So are the rows in both tables the same for these three primes. What 5, 13, and 17 have that the rest of the primes less than 29 do not is the property of being congruent to 1 (mod 4). On this evidence, we might make the correct guess:

If either p or q is congruent to 1 (mod 4), then $(p/q) = (q/p)$.

All of the entries not covered by this rule change sign from one table to the next. This behavior can be explained by the following hypothesis:

If p and q are both congruent to 3 (mod 4), then $(p/q) = -(q/p)$.

These guesses are in fact generally true, and they make up

Theorem 4. The Quadratic Reciprocity Theorem. If p and q are odd primes and $p \equiv q \equiv 3$ (mod 4), then $(p/q) = -(q/p)$. Otherwise, $(p/q) = (q/p)$.

We will postpone the proof of this theorem until the next section, but we will not hesitate to apply it for lack of a proof. Suppose that we want to see if $x^2 \equiv 85$ (mod 97) has a solution. That is, we want to evaluate $(85/97)$. With Theorems 3 and 4, we can carry the evaluation to a conclusion. We have

(7) $(85/97) = (17 \cdot 5/97) = (17/97)(5/97)$

by property (C) in Theorem 3. We will attack each factor in (7) separately. Because $97 \equiv 1$ (mod 4) (and, for that matter, $17 \equiv 1$ (mod 4) too), the quadratic reciprocity theorem says that

$$(17/97) = (97/17).$$

Property (A) in Theorem 3 says that

$$(97/17) = (12/17)$$

and

$$
\begin{aligned}
(12/17) = (4 \cdot 3/17) = (4/17)(3/17) \quad &\text{(by (C))} \\
= (3/17) \quad &\text{(by (B))} \\
= (17/3) \quad &\text{(by Theorem 4)} \\
= (2/3) \quad &\text{(by (A))} \\
= -1 \quad &\text{(by inspection).}
\end{aligned}
$$

The other factor is simpler:

$$(5/97) = (97/5) \qquad \text{(by Theorem 4)}$$
$$= (2/5) \qquad \text{(by (A))}$$
$$= -1 \qquad \text{(by inspection)}.$$

Putting these calculations back in (7), we get

$$(85/97) = (17/97)(5/97) = (-1)(-1) = 1;$$

thus the congruence has a solution. By applying (A) first, we could have evaluated (85/97) in another way:

$$(85/97) = (-12/97) = (-1/97)(4/97)(3/97) = (-1/97)(3/97).$$

We see that

$$(3/97) = (97/3) = (1/3) = 1,$$

so $(85/97) = (-1/97)$; if we know $(-1/97)$, we then know $(85/97)$.

If you look at a number of examples of Legendre symbols, it will become evident that to evaluate any one by using Theorems 3 and 4, it is enough to know what $(-1/p)$ and $(2/p)$ are for any p. Euler's Criterion quickly gives us $(-1/p)$:

Theorem 5. If p is an odd prime, then

$$(-1/p) = 1 \qquad \text{if} \qquad p \equiv 1 \ (\text{mod } 4),$$

and

$$(-1/p) = -1 \qquad \text{if} \qquad p \equiv 3 \ (\text{mod } 4).$$

In words, -1 is a quadratic residue of primes congruent to 1 (mod 4), and a nonresidue of all other odd primes.

Proof. Euler's Criterion says that

$$(-1/p) \equiv (-1)^{(p-1)/2} \ (\text{mod } p);$$

since $(p-1)/2$ is even if $p \equiv 1$ (mod 4) and $(p-1)/2$ is odd if $p \equiv 3$ (mod 4), the theorem is proved.

In the example we were just considering, $(-1/97) = 1$ because $97 \equiv 1$ (mod 4).

Theorem 5 tells us that we can sometimes find square roots of -1 modulo p: whenever $p \equiv 1 \pmod 4$, -1 has a square root $\pmod p$.

* **Exercise 14.** For which of the primes 3, 5, 7, 11, 13, 17, 19, and 23 is -1 a quadratic residue?

* **Exercise 15.** Evaluate (6/7) and (2/23)(11/23).

It is not so easy to determine whether 2 has a square root $\pmod p$. Euler's Criterion says that

$$(2/p) \equiv 2^{(p-1)/2} \pmod p,$$

but it is not obvious for which primes $2^{(p-1)/2}$ is congruent to 1 $\pmod p$. We will find out in the next section. For now, we state the result:

Theorem 6. If p is an odd prime, then

$$(2/p) = 1 \qquad \text{if} \qquad p \equiv 1 \qquad \text{or} \qquad 7 \pmod 8,$$
$$(2/p) = -1 \qquad \text{if} \qquad p \equiv 3 \qquad \text{or} \qquad 5 \pmod 8.$$

Theorem 6 together with Theorems 3 to 5 enables us to evaluate any Legendre symbol. For example, we can now evaluate (3201/8191). The calculations go as follows:

$$(3201/8191) = (3/8191)(11/8191)(97/8191);$$
$$(3/8191) = -(8191/3) = -(1/3) = -1,$$
$$(11/8191) = -(8191/11) = -(7/11) = (11/7) = (4/7) = 1,$$

and

$$(97/8191) = (8191/97) = (43/97) = (97/43) = (11/43)$$
$$= -(43/11) = -(-1/11) = 1.$$

Thus we see that $(3201/8191) = (-1)(1)(1) = -1$. Compare the labor of evaluating (3201/8191) as we have just done with that of determining by trial and error whether $x^2 \equiv 3201 \pmod{8191}$ has a solution. To calculate $1^2, 2^2, \ldots, 4095^2$ and divide each by 8191 is no light task. Theorems 3 to 6 are, for this job at least, an enormous help.

Problems

* 1. Which of the following congruences have solutions?

$$x^2 \equiv 7 \pmod{53} \qquad x^2 \equiv 14 \pmod{31}$$
$$x^2 \equiv 53 \pmod{7} \qquad x^2 \equiv 25 \pmod{997}$$

 2. Which of the following congruences have solutions?

$$x^2 \equiv 8 \pmod{53} \qquad x^2 \equiv 15 \pmod{31}$$
$$x^2 \equiv 54 \pmod{7} \qquad x^2 \equiv 625 \pmod{9973}$$

* 3. Find solutions for the congruences in Problem 1 that have them.

 4. Find solutions for the congruences in Problem 2 that have them.

* 5. Calculate $(33/71)$, $(34/71)$, $(35/71)$, and $(36/71)$.

 6. Calculate $(33/73)$, $(34/73)$, $(35/73)$, and $(36/73)$.

* 7. Solve $2x^2 + 3x + 1 \equiv 0 \pmod{7}$ and $2x^2 + 3x + 1 \equiv 0 \pmod{101}$.

 8. Solve $3x^2 + x + 8 \equiv 0 \pmod{11}$ and $3x^2 + x + 52 \equiv 0 \pmod{11}$.

* 9. Calculate $(1234/4567)$ and $(4321/4567)$.

 10. Calculate $(1356/2467)$ and $(6531/2467)$.

 11. Show that if $p = q + 4a$ (p and q are odd primes), then $(p/q) = (a/q)$.

 12. Show that if $p = 12k + 1$ for some k, then $(3/p) = 1$.

 13. Show that Theorem 6 could also be written $(2/p) = (-1)^{(p^2-1)/8}$ for odd primes p.

 14. Show that the quadratic reciprocity theorem could also be written $(p/q)(q/p) = (-1)^{(p-1)(q-1)/4}$ for odd primes p and q.

† 15. Student A says, "I've checked all the way up to 100 and I still haven't found n so that $n^2 + 1$ is divisible by 7. I'm tired now—I'll find one tomorrow." Student B says, after a few seconds of reflection, "No you won't." How did B know so quickly?

 16. Show that if a is a quadratic residue (mod p) and $ab \equiv 1 \pmod{p}$ then b is a quadratic residue (mod p).

* 17. Does $x^2 \equiv 211 \pmod{159}$ have a solution? Note that 159 is not prime.

 18. Prove that if $p \equiv 3 \pmod{8}$ and $(p - 1)/2$ is prime, then $(p - 1)/2$ is a quadratic residue (mod p).

* 19. Generalize Problem 16 by finding what condition on r will guarantee that if a is a quadratic residue (mod p) and $ab \equiv r \pmod{p}$, then b is a quadratic residue (mod p).

 20. Suppose that $p = q + 4a$, where p and q are odd primes. Show that $(a/p) = (a/q)$.

12

Quadratic Reciprocity

In this section we will prove the two theorems stated and used without proof in the last section: the quadratic reciprocity theorem and the theorem that enables us to evaluate $(2/p)$. The proof of the quadratic reciprocity theorem is not easy. Gauss guessed that the theorem was true in 1795, at the age of 18, after tabulating more than 10,000 values of (p/q), but he was not able to find his first proof of it for more than a year. Any proof that could elude the mighty mind of Gauss for that long is not easy to find.

At the base of both theorems is the following result, sometimes called *Gauss's Lemma:*

Theorem 1. Suppose that p is an odd prime, $p \nmid a$, and there are among the least residues (mod p) of

$$a, 2a, 3a, \ldots, \left(\frac{p-1}{2}\right) a$$

exactly g that are greater than $(p - 1)/2$. Then $x^2 \equiv a \pmod{p}$ has a solution or no solution according as g is even or odd. Otherwise stated,

$$(a/p) = (-1)^g.$$

Before proving the theorem, we will illustrate its application by taking $a = 5$ and $p = 17$. We have $(p - 1)/2 = 8$, and the integers

$$5, 10, 15, 20, 25, 30, 35, 40$$

have least residues (mod 17)

$$5, 10, 15, 3, 8, 13, 1, 6.$$

Three of these are greater than $(p - 1)/2$. Theorem 1 then says that 5 is a quadratic nonresidue (mod 17), which is so.

Exercise 1. Check that the theorem gives the right result in this case by applying Euler's Criterion and showing that $5^8 \equiv -1 \pmod{17}$.

Proof of Theorem 1. Let

$$r_1, r_2, \ldots, r_k$$

denote the least residues (mod p) of

$$a, 2a, \ldots, ((p - 1)/2)a$$

that are less than or equal to $(p - 1)/2$, and let

$$s_1, s_2, \ldots, s_g$$

denote those that are greater than $(p - 1)/2$. Thus $k + g = (p - 1)/2$. To prove the theorem, from Euler's Criterion it is enough to show that

$$a^{(p-1)/2} \equiv (-1)^g \pmod{p},$$

and this is what we proceed to do. In the example above, $k = 5$, and $g = 3$; the set of r's is $\{5, 3, 8, 1, 6\}$, and the set of s's is $\{10, 15, 13\}$. Both in the example and in general, no two of the r's are congruent (mod p). Suppose that two were. Then we would have for some k_1 and k_2,

$$k_1 a \equiv k_2 a \pmod{p}, \qquad 0 \le k_1 \le (p - 1)/2, \quad 0 \le k_2 \le (p - 1)/2.$$

Because $(a, p) = 1$, it follows that $k_1 = k_2$. For the same reason, no two of the s's are congruent (mod p). Now, consider the set of numbers

(1) $$r_1, r_2, \ldots, r_k, p - s_1, p - s_2, \ldots, p - s_g.$$

Each integer n in the set satisfies $1 \le n \le (p - 1)/2$, and there are $(p - 1)/2$ elements in the set. Gauss noticed, and we will now prove, that the numbers in the set are all different. From this it will follow that the elements in (1) are just a permutation of the integers

(2) $$1, 2, \ldots, (p - 1)/2,$$

and thus the product of the elements in (1) is the same as the product of the elements in (2). From this the theorem will follow. In the example we considered, the set of r's was $\{5, 3, 8, 1, 6\}$, and the set of $(p - s)$'s was $\{7, 2, 4\}$; between them, they include all the integers from 1 to 8.

To show that the elements in (1) are distinct, we have only to show that

$$r_i \not\equiv p - s_j \ (\text{mod } p)$$

for any i and j, because we have already seen that the r's and s's are distinct among themselves. Suppose that for some i and j we have

$$r_i \equiv p - s_j \ (\text{mod } p).$$

Then $r_i + s_j \equiv 0 \ (\text{mod } p)$. Since $r_i \equiv ta \ (\text{mod } p)$ and $s_j \equiv ua \ (\text{mod } p)$ for some t and u, with t and u positive integers less than or equal to $(p - 1)/2$, we would have

$$(t + u)a \equiv 0 \ (\text{mod } p);$$

since $(a, p) = 1$, we have $t + u \equiv 0 \ (\text{mod } p)$, and this is impossible, because $2 \leq t + u \leq p - 1$. Thus all of the elements in the set (1) are distinct, and consequently are a rearrangement of the elements in (2). Hence,

$$(3) \quad r_1 r_2 \cdots r_k (p - s_1)(p - s_2) \cdots (p - s_g) = 1 \cdot 2 \cdots ((p - 1)/2).$$

Because $p - s_j \equiv -s_j \ (\text{mod } p)$ for all j, and because there are g such terms, (3) becomes

$$(4) \qquad r_1 r_2 \cdots r_k s_1 s_2 \cdots s_g (-1)^g \equiv \left(\frac{p - 1}{2}\right)! \ (\text{mod } p).$$

But $r_1, r_2, \ldots, r_k, s_1, s_2, \ldots, s_g$ are, by definition, the least residues $(\text{mod } p)$ of

$$a, 2a, \ldots, ((p - 1)/2)a$$

in some order, so that the product $r_1 r_2 \cdots r_k s_1 s_2 \cdots s_g$ is congruent $(\text{mod } p)$ to $a(2a)(3a) \cdots ((p - 1)/2)a$. Thus (4) gives

$$a^{(p-1)/2}(-1)^g \left(\frac{p - 1}{2}\right)! \equiv \left(\frac{p - 1}{2}\right)! \ (\text{mod } p).$$

The common factor is relatively prime to p and may be canceled to give

$$a^{(p-1)/2}(-1)^g \equiv 1 \ (\text{mod } p).$$

If we multiply both sides of the last congruence by $(-1)^g$, we have

$$a^{(p-1)/2} \equiv (-1)^g \pmod{p}.$$

But we know that $a^{(p-1)/2} \equiv (a/p) \pmod{p}$. Putting the last two congruences together, and noting that if the two numbers are congruent (mod p), then they must be equal, we have

$$(a/p) = (-1)^g,$$

and this is what we wanted to prove.

* **Exercise 2.** Apply the theorem to determine whether $x^2 \equiv 7 \pmod{23}$ has a solution.

We will now apply the last theorem to evaluate $(2/p)$ for any odd prime p. According to the theorem, we need to find out how many of the least residues (mod p) of

(5) $$2, 4, 6, \ldots, 2\left(\frac{p-1}{2}\right)$$

are greater than $(p-1)/2$. Since all the numbers in (5) are already least residues, none of them being larger than p, we have only to see how many of them are greater than $(p-1)/2$. Let the first even integer greater than $(p-1)/2$ be $2a$. Between 2 and $(p-1)/2$ there are $a-1$ even integers, namely $2, 4, 6, \ldots, 2(a-1)$. So, the number of even integers from 2 to $p-1$ which are greater than $(p-1)/2$ is the total number of even integers, $(p-1)/2$, minus the number less than $(p-1)/2$, which is $a-1$. That is,

$$g = \frac{p-1}{2} - (a-1).$$

But

$2a$ is the smallest integer greater than $\dfrac{p-1}{2}$, so

a is the smallest integer greater than $\dfrac{p-1}{4}$, so

$a-1$ is the smallest integer greater than $\dfrac{p-5}{4}$, so

$-(a - 1)$ is the largest integer less than $-\left(\dfrac{p-5}{4}\right)$, so

g is the largest integer less than $\dfrac{p-1}{2} - \dfrac{p-5}{4}$:

g is thus the largest integer less than $(p + 3)/4$.

Exercise 3. Verify that the entries in the following table are correct.

p	3	5	7	11	13	17	19	23	29
g	1	1	2	3	3	4	5	6	7

Suppose that $p \equiv 1 \pmod 8$. Then $p = 1 + 8k$ for some k, and $(p + 3)/4 = (4 + 8k)/4 = 1 + 2k$. It follows that $g = 2k$ and that $(-1)^g = 1$. From Theorem 1, $(2/p) = 1$ if $p \equiv 1 \pmod 8$. Suppose that $p \equiv 3 \pmod 8$. Then $p = 3 + 8k$ for some k, and $(p + 3)/4 = (6 + 8k)/4 = 2k + 3/2$. It follows that $g = 2k + 1$ and that $(-1)^g = -1$. From Theorem 1, $(2/p) = -1$ if $p \equiv 3 \pmod 8$.

Exercise 4. Check the cases $p \equiv 5 \pmod 8$ and $p \equiv 7 \pmod 8$.

Thus we have proved

Theorem 2. If p is an odd prime, then

$$(2/p) = 1 \quad \text{if} \quad p \equiv 1 \quad \text{or} \quad 7 \pmod 8,$$
$$(2/p) = -1 \quad \text{if} \quad p \equiv 3 \quad \text{or} \quad 5 \pmod 8.$$

As an example of the use of Theorem 2, we will state and prove a result that is not used later, but is pleasing. Although we know when a number has primitive roots, finding the actual roots is generally not easy. For example, 2 is a primitive root of 3, 5, 11, 13, 19, 29, 37, 53, 59, 61, 67, and 83 among the primes less than 100, and it is not a primitive root of the others. No theorem has been proved that will tell which primes 2 is a primitive root of, and it has not even been proved that 2 is a primitive root of infinitely many primes. But we do have

Theorem 3. If p and $4p + 1$ are both primes, then 2 is a primitive root of $4p + 1$.

Proof. Let $q = 4p + 1$. Then $\phi(q) = 4p$, so 2 has order 1, 2, 4, p, $2p$, or $4p$ (mod q); we will show that the first five cases are impossible. We have

$$2^{2p} \equiv 2^{(q-1)} \equiv (2/q) \ (\text{mod } q)$$

by Euler's Criterion. But p is odd, so $4p \equiv 4$ (mod 8), and $q \equiv 4p + 1 \equiv 5$ (mod 8); we know from Theorem 2 that 2 is a quadratic non-residue of primes congruent to 5 (mod 8). Hence

$$2^{2p} \equiv -1 \ (\text{mod } q),$$

so 2 does not have order $2p$. Nor can the order of 2 be any of the divisors of $2p$, which are of course 1, 2, and p. Since 2 does not have order 4 either ($2^4 \equiv 1$ (mod q) implies $q \mid 15$, so $q = 5$, which is impossible), the theorem is proved.

We will now give Gauss's third proof of the quadratic reciprocity theorem. It depends on Gauss's Lemma (Theorem 1) and on a lemma we will now prove.

Lemma 1. If p and q are different odd primes, then

$$\sum_{k=1}^{(p-1)/2} \left[\frac{kq}{p} \right] + \sum_{k=1}^{(q-1)/2} \left[\frac{kp}{q} \right] = \frac{p-1}{2} \cdot \frac{q-1}{2}.$$

The notation $[kq/p]$ denotes the greatest integer not larger than kq/p. For example, take $p = 11$ and $q = 7$. Then

$$(6) \qquad \sum_{k=1}^{5} \left[\frac{7k}{11} \right] = \left[\frac{7}{11} \right] + \left[\frac{14}{11} \right] + \left[\frac{21}{11} \right] + \left[\frac{28}{35} \right] + \left[\frac{35}{11} \right]$$
$$= 0 + 1 + 1 + 2 + 3 = 7$$

and

$$(7) \qquad \sum_{k=1}^{3} \left[\frac{11k}{7} \right] = \left[\frac{11}{7} \right] + \left[\frac{22}{7} \right] + \left[\frac{33}{7} \right]$$
$$= 1 + 3 + 4 = 8,$$

and $(p - 1)(q - 1)/4 = 15 = 8 + 7$. The numbers have a geometrical interpretation: [35/11] is the number of positive integers less than 35/11, and in the figure below that is the number of *lattice points* (points with integer coefficients) above the *x*-axis and below the line $y = 7x/11$ when $x = 5$. The other terms in (6) are the number of

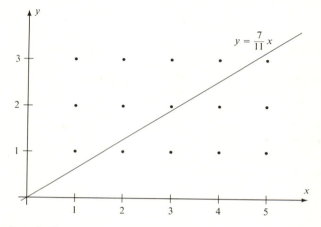

lattice points below the line when $x = 1$, 2, 3, and 4. In the same way, the terms in (7) are the number of lattice points to the left of the line and to the right of the *y*-axis for $y = 1$, 2, and 3. The number of lattice points in the 5-by-3 rectangle is 15.

Exercise 5. Verify that the lemma is true for $p = 5$ and $q = 7$.

Proof of Lemma 1. The idea used in the example works in general. Let

$$S(p, q) = \sum_{k=1}^{(p-1)/2} \left[\frac{kq}{p} \right] ;$$

we are thus trying to prove that

$$S(p, q) + S(q, p) = (p - 1)(q - 1)/4.$$

The figure on page 101 shows the same geometry as the figure on this page, but in general (actually, in the figure on page 101, $q = 13$ and $p = 11$). Just as in the example, $S(p, q)$ is the number of lattice points below the line $y = qx/p$ and above the *x*-axis for $x = 1, 2, \ldots ,$ $(p - 1)/2$. Also, $S(q, p)$ is the number of lattice points to the left of the line and to the right of the *y*-axis. There are no lattice points on the line, because if (a, b) were on the line, then $b = qa/p$ or $bp = qa$, and this is impossible. (Since $p \mid bp$, we have $p \mid qa$ and since $(p, q) = 1$, it follows

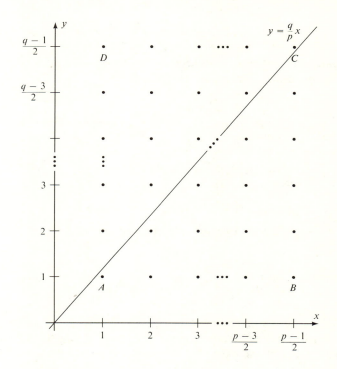

that $p \mid a$. But $1 \leq a \leq (p-1)/2$, and there are no multiples of p in that interval.) Thus each of the lattice points in or on the boundary of the rectangle $ABCD$ is counted exactly once; on the one hand, this number is $S(p, q) + S(q, p)$, and on the other it is $((p-1)/2)((q-1)/2)$. This proves the lemma.

Theorem 4. The Quadratic Reciprocity Theorem. If p and q are odd primes, then

$$(p/q)(q/p) = (-1)^{(p-1)(q-1)/4}.$$

Note that this is equivalent to the way we stated the theorem earlier (Theorem 4 of Section 11): If $p \equiv q \equiv 3 \pmod 4$, then $(p/q) = -(q/p)$; otherwise, $(p/q) = (q/p)$. This is so because $(p-1)(q-1)/4$ is even unless $p \equiv q \equiv 3 \pmod 4$.

Proof. As in the proof of Gauss's Lemma, let us take the least residues (mod p) of

$$q, 2q, 3q, \ldots, \frac{p-1}{2} q$$

and separate them into two classes. Put those less than or equal to $(p-1)/2$ in one class and call them

$$r_1, r_2, \ldots, r_k,$$

and put those greater than $(p-1)/2$ in another and call them

$$s_1, s_2, \ldots, s_g.$$

Thus $k + g = (p-1)/2$. The conclusion of Gauss's Lemma was that $(q/p) = (-1)^g$. For short, let

$$R = r_1 + r_2 + \cdots + r_k \qquad \text{and} \qquad S = s_1 + s_2 + \cdots + s_g.$$

While proving Gauss's Lemma, we showed that the set of numbers

(8) $r_1, r_2, \ldots, r_k, p - s_1, p - s_2, \ldots, p - s_g$

was a permutation of

(9) $1, 2, \ldots, (p-1)/2.$

It follows that the sum of the elements in (8) is the same as the sum of the elements in (9). Remember that in the proof of Gauss's Lemma, we took the *product* of the elements in (8) and equated it to the product of the elements in (9). Here, then, is a possible starting point of this proof: Gauss may have thought (in German), "What would happen if I equated the sums instead of the products?" and then constructed the proof. Whatever it was that he thought, the lesson to be learned is that proofs often do not start at the beginning. The sum of the numbers in (9) is, by a well-known formula $(1 + 2 + \cdots + n = n(n+1)/2$; see Appendix A for a proof),

$$\frac{1}{2}\left(\frac{p-1}{2}\right)\left(\frac{p-1}{2} + 1\right) = \frac{p^2 - 1}{8}.$$

The sum of the elements in (8) is

$$\sum_{j=1}^{k} r_j + \sum_{j=1}^{g} (p - s_j) = R + gp - S.$$

Thus we have

(10) $R = S - gp + (p^2 - 1)/8.$

The least residue (mod p) of jq $(j = 1, 2, \ldots, (p-1)/2)$ is the remainder when we divide jq by p. We know the quotient, $[jq/p]$, so if we let t_j denote the least residue (mod p) of jq, we have

$$jq = [jq/p]p + t_j,$$

$j = 1, 2, \ldots , (p - 1)/2$. If we sum these equations over j, we have

$$\sum_{j=1}^{(p-1)/2} jq = \sum_{j=1}^{(p-1)/2} [jq/p]p + \sum_{j=1}^{(p-1)/2} t_j,$$

or

$$q \sum_{j=1}^{(p-1)/2} j = p \sum_{j=1}^{(p-1)/2} [jq/p] + \sum_{j=1}^{k} r_j + \sum_{j=1}^{g} s_j,$$

or

(11) $$q(p^2 - 1)/8 = pS(p, q) + R + S.$$

Substituting into this from (10), we get

$$q(p^2 - 1)/8 = pS(p, q) + 2S - gp + (p^2 - 1)/8,$$

or

(12) $$(q - 1)(p^2 - 1)/8 = p(S(p, q) - g) + 2S.$$

In (12), the left-hand side is even (because $(p^2 - 1)/8$ is an integer and $q - 1$ is even), and $2S$ is even. It follows that the remaining term in (12) is even, and so $S(p, q) - g$ is even. Hence

$$(-1)^{S(p,q)-g} = 1.$$

Since $(-1)^g = (q/p)$, then

(13) $$(-1)^{S(p,q)} = (-1)^g = (q/p).$$

Now we can repeat the argument with p and q interchanged—nowhere have we required q to have a property that p does not—and get

(14) $$(-1)^{S(q,p)} = (p/q).$$

Multiplying (13) and (14), we have

$$(-1)^{S(p,q)+S(q,p)} = (p/q)(q/p),$$

and from Lemma 1, we have

$$(-1)^{(p-1)(q-1)/4} = (p/q)(q/p),$$

which is what we wanted.

The proof may seem to be unsatisfying: each step is correct, and hence the conclusion is true, but it is not clear why the steps are there and where they came from. That is because there are at least eight

levels of mathematical understanding, and it is hard for someone on a lower level to appreciate what goes on at a higher level. The levels are, I think:

1. Being able to do arithmetic.
2. Being able to substitute numbers in formulas.
3. Given formulas, being able to get other formulas.
4. Being able to understand the hypotheses and conclusions of theorems.
5. Being able to understand the proofs of theorems, step by step.
6. Being able to *really* understand the proofs of theorems: that is, seeing why the proof is as it is, and comprehending the inwardness of the theorem and its relation to other theorems.
7. Being able to generalize and extend theorems.
8. Being able to see new relationships and discover and prove entirely new theorems.

Those of us stuck at level 5 can no more understand the workings of a level 8 mind than a cow could understand calculus.

Problems

*† 1. Adapt the method used in the text to evaluate $(2/p)$ to evaluate $(3/p)$.

2. Show that 3 is a quadratic nonresidue of all primes of the form $4^n + 1$.

† 3. Show that 3 is a quadratic nonresidue of all Mersenne primes greater than 3.

4. (a) Prove that if $p \equiv 7 \pmod{8}$, then $p \mid (2^{(p-1)/2} - 1)$.
 (b) Find a factor of $2^{83} - 1$.

† 5. (a) If p and $q = 10p + 3$ are odd primes, show that $(p/q) = (3/p)$.
 (b) If p and $q = 10p + 1$ are odd primes, show that $(p/q) = (-1/p)$.

6. (a) Which primes can divide $n^2 + 1$ for some n?
 (b) Which odd primes can divide $n^2 + n$ for some n?
 (c) Which odd primes can divide $n^2 + 2n + 2$ for some n?

7. (a) Show that if $p \equiv 3 \pmod{4}$ and a is quadratic residue \pmod{p}, then $p - a$ is a quadratic nonresidue \pmod{p}.
 * (b) What if $p \equiv 1 \pmod{4}$?

8. If $p > 3$, show that p divides the sum of its quadratic residues that are also least residues.

*✝ 9. If p is an odd prime, evaluate

$$(1 \cdot 2/p) + (2 \cdot 3/p) + \cdots + ((p - 2)(p - 1)/p).$$

10. Show that if $p \equiv 1 \pmod 4$, then $x^2 \equiv -1 \pmod p$ has a solution given by the least residue $\pmod p$ of $((p - 1)/2)!$.

13

Numbers in Other Bases

One of the great accomplishments of the human mind, and one that made mathematics possible, was the invention of our familiar notation for writing integers. We write integers in a place-value notation, with each place indicating a different power of 10. For example,

$$314{,}159 = 3 \cdot 10^5 + 1 \cdot 10^4 + 4 \cdot 10^3 + 1 \cdot 10^2 + 5 \cdot 10^1 + 9 \cdot 10^0.$$

The first people to use place-value notation were the Babylonians of more than 3000 years ago. None of the other ancient civilizations—not the Egyptians, the Chinese, or the Greeks—had the place-value idea, and if the Babylonians had not discovered it, it might have remained forever undiscovered, with great consequences for the history of the human race. Among other things, this book would never have been written. The Babylonians transmitted the idea to the Hindus (before 600 BC), from whom the Arabs got it (by 600 AD), and the Arabs made it known in Europe (1200 AD).

There is no reason why some integer other than 10 could not be used for the same purpose. The choice of 10 is only an anatomical accident. In fact, other integers—we will call them *bases*—have been used in the past. The Babylonians sometimes used the base 60, and the ancient Mayans used the base 20. Today, numbers written in the bases 2, 8, and

16 are used by computers. In this section, we will look at integers in bases other than 10.

We start by looking at a special case.

Theorem 1. Every positive integer can be written as a sum of distinct powers of two.

For example, $22 = 2^4 + 2^2 + 2^1$ and $23 = 2^4 + 2^2 + 2^1 + 2^0$. But $24 = 2^3 + 2^3 + 2^3$ is not a proper representation, because the powers of two are not distinct.

* **Exercise 1.** Write 31 and 33 as sums of distinct powers of two.

Proof of Theorem 1. The idea of the proof is to take an integer n and subtract from it the largest power of 2 that is smaller than it—say 2^k. Then we do the same for $n - 2^k$. If we continue this process, we will eventually get a representation of n in the form that we want. To make this argument rigorous, we prove the theorem by induction: $1 = 2^0$, $2 = 2^1$, and $3 = 2^0 + 2^1$, so the theorem is true if the integer is 1, 2, or 3. Suppose now that every integer k, $k \leq n - 1$, can be written as a sum of distinct powers of 2. We want to show that n can also be so written. We know that n falls between some two distinct powers of 2; that is, there is an integer r such that

$$2^r \leq n < 2^{r+1}.$$

* **Exercise 2.** What is r if $n = 74$? If $n = 174$?

The largest power of 2 not larger than n is 2^r. Let $n' = n - 2^r$. Then $n' \leq n - 1$, so the induction assumption tells us that it can be written as the sum of distinct powers of two:

$$n' = 2^{e_1} + 2^{e_2} + \cdots + 2^{e_k},$$

where $e_i \neq e_j$ if $i \neq j$. Since $n' = n - 2^r$, we have

(1) $$n = 2^r + 2^{e_1} + 2^{e_2} + \cdots + 2^{e_k},$$

and so n can be written as a sum of powers of two. To complete the proof, we need to show that r is different from any of e_1, e_2, \ldots, e_k.

* **Exercise 3.** Show that this is true.

We now show that the representation (1) is unique.

Theorem 2. Every positive integer can be written as the sum of distinct powers of 2 in only one way.

Proof. Suppose that n has two representations as a sum of distinct powers of 2. We will show that the representations are really the same. To make the notation less clumsy, we note that any sum of distinct powers of 2 can be written in the form

$$(2) \qquad d_0 + d_1 \cdot 2 + d_2 \cdot 2^2 + d_3 \cdot 2^3 + \cdots + d_k \cdot 2^k$$

for some k, where $d_i = 0$ or 1 for each i. Conversely, every such sum is a sum of distinct powers of 2. Hence it is immaterial whether we write n in the form (1) or (2), and (2) has the advantage of having no subscripts on the exponents. If n has two representations, we have

$$(3) \qquad \begin{aligned} n &= d_0 + d_1 \cdot 2 + d_2 \cdot 2^2 + \cdots + d_k \cdot 2^k \\ &= e_0 + e_1 \cdot 2 + e_2 \cdot 2^2 + \cdots + e_k \cdot 2^k, \end{aligned}$$

where $d_i = 0$ or 1 and $e_i = 0$ or 1 for each i. (Note that we lose no generality in assuming that the two representations have the same number of terms. If one is longer than the other, we can add zero terms to the shorter one until the two have the same length.) Subtracting the second representation in (3) from the first gives

$$(4) \quad 0 = (d_0 - e_0) + (d_1 - e_1) \cdot 2 + (d_2 - e_2) \cdot 2^2 + \cdots + (d_k - e_k) \cdot 2^k$$

Hence $2 \mid (d_0 - e_0)$. But since d_0 and e_0 are either 0 or 1, it follows that

$$-1 \leq d_0 - e_0 \leq 1.$$

Since the only multiple of 2 in that range is zero, $d_0 = e_0$. Thus the first term in (4) disappears, and we may divide what remains by 2 to get

$$(5) \qquad 0 = (d_1 - e_1) + (d_2 - e_2) \cdot 2 + \cdots + (d_k - e_k) \cdot 2^{k-1}.$$

The same argument as before shows that $d_1 = e_1$. Dropping the term $d_1 - e_1$ from (5), dividing by 2 and applying the same argument again, we get $d_2 = e_2$. And so on: $d_3 - e_3 = d_4 - e_4 = \cdots = d_k - e_k = 0$, and the two representations in (5) were the same.

Theorems 1 and 2 show that every n can be written in exactly one way in the form

(6) $$d_0 + d_1 \cdot 2 + d_2 \cdot 2^2 + d_3 \cdot 2^3 + \cdots + d_k \cdot 2^k$$

for some k, where each d is either 0 or 1. This is like the ordinary decimal representation for integers. It is so like it that we can write numbers of the form (6) in the same style as we usually write integers. The powers of 2 and the plus signs in (6) are not essential, since the thing that determines n is the sequence d_0, d_1, \ldots, d_k. We will write the expression (6) as

$$(d_k d_{k-1} \cdots d_1 d_0)_2,$$

and say that we have written the integer *in the base* 2. The subscript 2 reminds us that d_r is multiplied by 2^r. For example,

$$101001_2 = 1 + 0 \cdot 2 + 0 \cdot 2^2 + 1 \cdot 2^3 + 0 \cdot 2^4 + 1 \cdot 2^5$$
$$= 1 + 8 + 32 = 41.$$

In the other direction,

$$94 = 64 + 16 + 8 + 4 + 2$$
$$= 1 \cdot 2^6 + 0 \cdot 2^5 + 1 \cdot 2^4 + 1 \cdot 2^3 + 1 \cdot 2^2 + 1 \cdot 2^1 + 0 \cdot 2^0$$
$$= 1011110_2.$$

* **Exercise 4.** Evaluate 1001_2, 111_2, and 1000000_2.

* **Exercise 5.** Write 2, 20, and 200 in the base 2.

We know that every integer can be uniquely expressed in the form

$$d_0 + d_1 \cdot 10 + d_2 \cdot 10^2 + \cdots + d_k \cdot 10^k$$

for some k, with $0 \le d_i < 10$, $i = 0, 1, \ldots, k$, and Theorems 1 and 2 show that every integer can be uniquely expressed in the form

$$d_0 + d_1 \cdot 2 + d_2 \cdot 2^2 + \cdots + d_k \cdot 2^k$$

for some k, with $0 \le d_i < 2$, $i = 0, 1, \ldots, k$. What we can do for 2 and 10, we ought to be able to do for any integer greater than 1. In fact, we can. We will prove

Theorem 3. Let $b \ge 2$ be any integer (called the base). Any positive integer can be written uniquely in the base b; that is, in the form

$$n = d_0 + d_1 \cdot b + d_2 \cdot b^2 + \cdots + d_k \cdot b^k$$

for some k, with $0 \le d_i < b$, $i = 0, 1, \ldots, k$.

Proof. We will first show that each integer has such a representation and then show that it is unique. To show that there is a representation, we could adapt the proof of Theorem 1, but the proof that we will present here (which could also be applied to Theorem 1) gives a construction for the digits of n in the base b. Divide n by b: the division algorithm says

$$n = q_1 b + d_0, \qquad 0 \le d_0 < b.$$

We can divide the quotient by b,

$$q_1 = q_2 b + d_1, \qquad 0 \le d_1 < b,$$

and continue the process,

$$q_2 = q_3 b + d_2, \qquad 0 \le d_2 < b,$$
$$q_3 = q_4 b + d_3, \qquad 0 \le d_3 < b,$$

and so on. Since $n > q_1 > q_2 > q_3 > \cdots$ and each q_i is nonnegative, the sequence of q's will sooner or later terminate. That is, we will come to k such that

$$q_k = 0 \cdot b + d_k \qquad 0 \le d_k < b.$$

But then

$$n = d_0 + q_1 b = d_0 + (d_1 + q_2 b)b = d_0 + d_1 b + q_2 b^2$$
$$= d_0 + d_1 b + (d_2 + q_3 b)b^2 = d_0 + d_1 b + d_2 b^2 + q_3 b^3$$
$$\cdots$$
$$= d_0 + d_1 b + d_2 b^2 + \cdots + d_{k-1} b^{k-1} + q_k b^k$$
$$= d_0 + d_1 b + d_2 b^2 + \cdots + d_k b^k,$$

and this is the desired representation.

To show that it is unique, we use the same idea we used in the proof of Theorem 2. Suppose that n has two representations:

$$n = d_0 + d_1 b + d_2 b^2 + \cdots + d_k b^k$$
$$= e_0 + e_1 b + e_2 b^2 + \cdots + e_k b^k$$

for some k, where

$$(7) \qquad 0 \le d_i < b \qquad \text{and} \qquad 0 \le e_i < b$$

for $i = 0, 1, \ldots, k$. (As in Theorem 2, there is no loss in generality in assuming that the two representations have the same number of terms.) Subtracting one representation from the other gives

$$0 = (d_0 - e_0) + (d_1 - e_1)b + (d_2 - e_2)b^2 + \cdots + (d_k - e_k)b^k.$$

We see that $b \mid (d_0 - e_0)$. From (7) it follows that $d_0 = e_0$.

Exercise 6. Complete the proof.

For short, we will write

$$d_0 + d_1 b + \cdots + d_k b^k = (d_k d_{k-1} \cdots d_1 d_0)_b.$$

For example,

$$111_7 = 1 + 1 \cdot 7 + 1 \cdot 7^2 = 57_{10}.$$

We will usually omit the subscript b when $b = 10$. Unless noted otherwise, every integer without a subscript is written in base 10.

To find the representation of a base-10 integer in the base b, the scheme used in the proof of Theorem 3 is as good as any. For example, to write 31415 in the base 8, we perform repeated divisions by 8:

$$31415 = 8 \cdot 3926 + 7,$$
$$3926 = 8 \cdot 490 + 6,$$
$$490 = 8 \cdot 61 + 2,$$
$$61 = 8 \cdot 7 + 5,$$
$$7 = 8 \cdot 0 + 7,$$

and hence $31415_{10} = 75267_8$. (Check:

$$75267_8 = 7 + 6 \cdot 8 + 2 \cdot 8^2 + 5 \cdot 8^3 + 7 \cdot 8^4 = 7 + 48 + 128 + 2560 + 28672$$
$$= 31415.)$$

To make the arithmetic easier, the divisions may be arranged differently. For example, we have $31415_{10} = 160406_7$:

quotients	remainder
7) 31415	
4487	6
641	0
91	4
13	0
1	6
0	1

Problems

* 1. Write 1492 in base

$$2 \quad 3 \quad 7 \quad 9 \quad 11.$$

2. Write 1776 in base

$$4 \quad 5 \quad 6 \quad 8 \quad 11.$$

* 3. Write in base 10:

$$3141_5 \quad 3141_7 \quad 3141_{11} \quad 3141_{12}.$$

4. Write in base 10:

$$1215_6 \quad 1215_8 \quad 1215_9 \quad 1215_{20}.$$

* 5. Solve for x:

$$123_4 = x_5 \quad 234_5 = x_6 \quad 123_x = 1002_4.$$

6. Solve for x:

$$345_6 = x_7 \quad 456_x = 2201_7 \quad 2x3_4 = 1x10_3.$$

* 7. Construct a multiplication table in base 7.

8. Construct a multiplication table in base 8.

* 9. All numbers in this problem are in base 9. Calculate, in base 9 (that is, no conversions to any other base):

$$15 + 24 + 33 \qquad 1620 - 1453$$
$$42 \cdot 12 \qquad\qquad 314 \cdot 152.$$

10. With the same instructions as in Problem 9, calculate

$$16 + 35 + 44 \qquad 1453 - 1066$$
$$53 \cdot 23 \qquad\qquad 425 \cdot 263.$$

* 11. Let $(.d_1 d_2 d_3 \cdots)_b$ stand for $d_1/b + d_2/b^2 + d_3/b^3 + \ldots$. Evaluate as rational numbers in base 10:

$$(.25)_7 \quad (.333\ldots)_7 \quad (.5454\ldots)_7.$$

12. With the same instructions as in Problem 11, evaluate

$$(.36)_8 \quad (.444\ldots)_8 \quad (.6565\ldots)_8.$$

* 13. In which bases b is 1111_b divisible by 5?

14. (a) Show that 123_7, 132_7, 312_7, 231_7, and 213_7 are even integers.
 (b) Show that in the base 7, an integer is even if and only if the sum of its digits is even.
 (c) In which other bases is it true that if an integer is even, then any permutation of its digits is even?

15. An eccentric philanthropist undertakes to give away $100,000. He is eccentric because he insists that each of his gifts be a number of dollars that is a power of two, and he will give no more than one gift of any amount. How does he distribute the money?

16. Prove that every positive integer can be written unique in the form

$$n = e_0 + 3e_1 + 3^2 e_2 + \cdots + 3^k e_k$$

for some k, where $e_i = -1, 0,$ or $1, i = 0, 1, \ldots, k$.

* 17. Hexadecimal notation (base 16) uses the digits A, B, C, D, E, and F for decimals 10, 11, 12, 13, 14, and 15.
 (a) Convert 3073, 53456, 49370, and 45278 to hexadecimal.
 (b) Convert CAB, B0B0, DEAF, and A1DE to decimal.
 (c) Is there a longer hexadecimal word than DEFACADED?

18. To convert from decimal to binary, it is convenient to convert first to octal, and then replace each octal digit with its binary representation. For example, $1929_{10} = 3611_8 = 11,110,001,001_2$ and $10,111,010,100_2 = 2724_8 = 1492_{10}$. Show that this process works in general.

19. Another method of representing integers is in the factorial notation:
 $(d_k d_{k-1} \cdots d_1)_! = d_1 \cdot 1! + d_2 \cdot 2! + \cdots + d_k \cdot k!, \qquad 0 \le d_i \le i$.
 * (a) Write $(22110)_!$ and $(242120)_!$ in base 10.
 * (b) Write 920 and 1848 in factorial notation.
 † (c) Prove that every positive integer has a unique representation in factorial notation.

20. Find a base b in which 45_b and 55_b are squares of consecutive integers.

Duodecimals

In this section, which has no new mathematical content and can be omitted without serious loss, we will take a close look at arithmetic in base 12. In doing this, we will reexamine the familiar processes of addition, subtraction, multiplication, and division in an unfamiliar setting. After completing the exercises, you should realize how expert you really are at computation (in the base 10), and what enormous labor it takes to become quick at arithmetical operations. We all worked hard in the third grade. Any base would serve as well as 12 to give practice, but some parts of arithmetic—notably decimals—are nicer in base 12 than they are in base 10. Besides, there is a good deal of twelveness in everyday life: items are measured by the dozen and gross, there are 12 months in the year, 12 inches in a foot, half a dozen feet in a fathom, two dozen hours in the day, and 30 dozen degrees in the circle. The reason for this abundance of twelves is the easy divisibility of 12 by 3, 4, and 6; we want to make such divisions much more often than we want to divide things by 5. Counting by tens is the result of a really unfortunate accident. How much better ordered the world would be if we had six fingers on each hand! Because we don't, it is unlikely that we will ever abandon counting by tens, even though counting by dozens is manifestly better. But there is a Duodecimal Society of America that

devotes itself to educating the public in preparation for the day when the change is made. Although the public is still largely untouched, the Society looks forward to that day with faint but unquenchable hope.

In order to count by dozens, we need two new digits to represent 10_{10} and 11_{10}. From now on in this section, all numbers will be *duodecimals*—that is, written in base 12, unless otherwise indicated with a subscript. The settled notation among duodecimalists for 10_{10} and 11_{10} seems to be χ and ϵ, the Greek letters *chi* and *epsilon*. But they are pronounced "dec" and "el." Thus duodecimal counting goes

$$1, 2, 3, 4, 5, 6, 7, 8, 9, \chi, \epsilon, 10, 11, \ldots, 1\chi, 1\epsilon, 20, \ldots,$$
$$30, \ldots, 40, \ldots, \chi0, \ldots, \epsilon0, \ldots, 100, \ldots.$$

We need names for duodecimal numbers. The Duodecimal Society advocates "do" (from dozen) for 10 and "gro" for 100, so that, for example, 15 is "do five" and 327 is "3 gro 2 do 7." Unfortunately, the Society's names for larger multiples of a dozen do not fall naturally from the lips: 1000, 10000, 100000, and 1000000 are, respectively, mo, do-mo, gro-mo, and bi-mo, so that 220110 would have to be read "2 bi-mo, 2 gro-mo, mo, gro, do." But if you want to call 1χ and 5ϵ "decteen" and "fifty-el" instead of "do dec" and "five do el," probably only the most fanatical duodecimalists would object.

Addition of duodecimals is not hard if we remember to carry one whenever we sum to a dozen. Here is the addition table for six:

6	6	6	6	6	6	6	6	6	6	6	6
1	2	3	4	5	6	7	8	9	χ	ϵ	10
7	8	9	χ	ϵ	10	11	12	13	14	15	16

and here are some summations:

5	31	123	$\chi\chi\chi$
4	41	456	$\epsilon\epsilon\epsilon$
3	15	789	$\chi\epsilon\chi$
2	9	$\chi\epsilon0$	$\epsilon\chi\epsilon$
12	94	2036	3996

* **Exercise 1.** Calculate $9 + 4$, $\chi + \epsilon$, $\epsilon1 + 1\epsilon$, and $16 + 19 + 37$.

Although we can carry the addition tables in our heads without too much trouble, we need a multiplication table to look at when we multiply, just as when we learned to multiply in the base χ.

Duodecimal Multiplication Table

·	2	3	4	5	6	7	8	9	χ	ε	10
2	4	6	8	χ	10	12	14	16	18	1χ	20
3	6	9	10	13	16	19	20	23	26	29	30
4	8	10	14	18	20	24	28	30	34	38	40
5	χ	13	18	21	26	2ε	34	39	42	47	50
6	10	16	20	26	30	36	40	46	50	56	60
7	12	19	24	2ε	36	41	48	53	5χ	65	70
8	14	20	28	34	40	48	54	60	68	74	80
9	16	23	30	39	46	53	60	69	76	83	90
χ	18	26	34	42	50	5χ	68	76	84	92	χ0
ε	1χ	29	38	47	56	65	74	83	92	χ1	ε0

Exercise 2. Verify that the χ-times table is correct.

With the aid of the table, multiplication is no problem. For example,

```
  34          1755         χχ
   5             χ         εε
 ---         ------        ---
 148         14262         9ε2
                          9ε2
                         -----
                         χ912
```

* **Exercise 3.** Calculate 14·2, 14·3, and 9·χ·ε.

With enough practice, we could absorb the entries of the table and learn to do without it; eventually, "9 times 9" would produce "6 do 9" purely by reflex.

Division is slightly harder, even with the aid of the table, because lack of experience may lead us to choose the wrong digit in a quotient. It takes practice to be able to see at a glance how many χ5's there are in 763.

* **Exercise 4.** How many are there?

Here are some worked-out divisions:

```
5 ) 456 ( χ8      22 ) 456 ( 20      31 ) 4159 ( 140
    42                 44                 31
    --                 --                 ---
    36                 16                 105
    34                                    104
    --                                    ---
     2                                     19
```

* **Exercise 5.** Calculate 1966/6 and 1111/5.

In duodecimals, 1/3 and 1/6 have terminating expansions—1/3 = 4/10 = .4 and 1/6 = 2/10 = .2—and this is pleasanter than the case in base χ. The expansion of 1/5, however, does *not* terminate: long division shows that 1/5 = .24972497 We will signify the repeating part of a repeating decimal by putting a bar over it; thus we will write 1/5 = .$\overline{2497}$. We will continue to call such things "decimals," even though some other name, perhaps "dozinals," would be more appropriate.

* **Exercise 6.** Calculate the decimal representation of 1/7.

Here is a table of reciprocals up through one decth and one elth.

n	2	3	4	5	6	7	8	9	χ	ϵ
1/n	.6	.4	.3	.$\overline{2497}$.2	.$\overline{186\chi35}$.16	.14	.$\overline{12497}$.$\overline{1}$

Half of these terminate, and one elth, like one-ninth in base χ, has a particularly simple repeating part.

To convert from decimals to rational numbers, we use the same principles as in base χ. For example, .25 = 25/100. This is a fraction in lowest terms, though your reflexes may tell you differently; actually, the numerator is a prime. Repeating decimals can be converted to fractions in the usual way. For example, let $N = .6666$ Then $10N = 6.6666$. . . , so $10N - N = 6$. Thus $\epsilon N = 6$, and $N = 6/\epsilon$; this is a fraction in its lowest terms.

Problems

* 1. Calculate 3141 + 5926 and 3141 · 5926 without changing to another base.
 2. Calculate 3141/5926 to three places without changing to another base.
* 3. Write 7/13 as a decimal and .2929. . . as a rational number.
 4. Write 8/14 as a decimal and .3030. . . as a rational number.
* 5. Which would you rather have, 4\epsilon$.$\epsilon6 or $(59.95)_\chi$?
 6. Show that the last digit of a square is 0, 1, 4, or 9.
 7. Show that the last digit of x^n, $n = 2, 3, \ldots$
 (a) is 0 if $x = 6$
 (b) is 4 if $x = \chi$
 (c) is x if $x = 3, 5, 7, 8,$ or ϵ and n is odd.

8. Show that any integer whose last digit is 3, 6, 9, or 0 has 3 for a factor.

9. Show that any integer whose last digit is 4, 8, or 0 has 4 for a factor, and that any integer whose last digit is 6 or 0 has 6 for a factor.

χ. Show that any integer whose digits sum to a multiple of ϵ is divisible by ϵ.

*† ϵ. The Duodecimal Society of America also advocates the do-metric system of weights and measures: 1000 yards to the mile, 10 ounces to the pound, and 10 flounces to the pint. The Society relates distance, weight, and volume by requiring that a cubic yard hold 1000 pints of water, which weighs 1000 pounds. If we keep the yard as it is now, how do the do-metric mile, pint, and pound compare to the ordinary mile, pint, and pound?

10. (a) How many days are there in the year?
 (b) What other three-digit numbers have the same property—that is, $d_1 d_2 d_3 = ((d_1 + 1)d_2 d_3)_x$?
 (c) Are there any four-digit numbers with this property?

Decimals

Decimals are very handy for calculation. If asked to add 2/5, 1/3, 3/10, 4/7, 5/9, and 3/8 together, any sensible calculator would change the fractions to decimals first. Decimals are so handy that the metric system, which is based on decimals, is replacing our traditional system of measurement, which is partly decimal, partly binary, partly duodecimal, and partly arbitrary.

Decimals seem so natural to us that it is easy to forget that they were a human invention and not something that was handed down to us from above, engraved on stone tablets. Our present system of notation, which we take for granted, was arrived at only after a long evolution, still going on recently. The first author whose works we could read today without being puzzled by the notation is Descartes, who wrote in the first half of the seventeenth century, and even he wrote xx instead of x^2 and \propto for $=$. The great period for evolution of notation was the sixteenth century, when each author of a book would use his own personal notation (there was no traditional notation, since printing had been invented only the century before). The fittest notations survived, and decimals, invented by Simon Stevin in the last half of the 1500's, was one of them.

Some fractions have simple decimal expansions ($1/8 = .125$); others

are not as simple, but still tolerable (1/3 = .333 . . .); and others are not simple at all (1/17 = .0588235294117647 . . .). In this section we will see which fractions are simplest, and we will find a way of determining how long the repeating part of a repeating decimal is without having to carry out the actual calculation of the decimal. In doing so, we will use nothing deeper than the division algorithm and some congruences.

We will denote

$$d_1/10 + d_2/10^2 + d_3/10^3 + \cdots \qquad (0 \le d_k < 10)$$

by

$$.d_1 d_2 d_3 \ . \ . \ . \ .$$

A bar over part of a decimal will indicate that this part repeats indefinitely. For example,

$$.01\overline{47} = .0147474747 \ . \ . \ . \qquad \text{and} \qquad 1/3 = .\overline{3}.$$

* **Exercise 1.** Write $.01\overline{47}$ as a rational number.

* **Exercise 2.** Write 7/41 as a decimal with a bar over its repeating part.

Let us make a table of the decimal expansions of the reciprocals of the first few integers and see if we can notice any pattern in them. A zero in the period column means that the decimal terminates.

n	$1/n$	Period	n	$1/n$	Period
2	.5	0	16	.0625	0
3	.$\overline{3}$	1	17	.$\overline{0588235294117647}$	16
4	.25	0	18	.0$\overline{5}$	1
5	.2	0	19	.$\overline{052631578947368421}$	18
6	.1$\overline{6}$	1	20	.05	0
7	.$\overline{142857}$	6	21	.0$\overline{47619}$	6
8	.125	0	22	.0$\overline{45}$	2
9	.$\overline{1}$	1	23	.$\overline{0434782608695652173913}$	22
10	.1	0	24	.041$\overline{6}$	1
11	.$\overline{09}$	2	25	.04	0
12	.08$\overline{3}$	1	26	.0$\overline{384615}$	6
13	.$\overline{076923}$	6	27	.$\overline{037}$	3
14	.0$\overline{714285}$	6	28	.03$\overline{571428}$	6
15	.0$\overline{6}$	1	29	.$\overline{0344827586206896551724137931}$	28

The integers in the table whose decimal reciprocals terminate are

$$2, 4, 5, 8, 10, 16, 20, 25,$$

and one thing these numbers have in common is that they are all of the form $2^a 5^b$ for some nonnegative integers a and b. We might guess that the decimal expansion of the reciprocal of any number of this form terminates. The next three such numbers are 32, 40, and 50, and

$$1/32 = .03125, \qquad 1/40 = .025, \qquad 1/50 = .02$$

all terminate. In fact, this guess is right.

Theorem 1. If a and b are any nonnegative integers, then the decimal expansion of $1/2^a 5^b$ terminates.

Proof. Let M be the maximum of a and b. Then

$$10^M (1/2^a 5^b) = 2^{M-a} 5^{M-b}$$

is an integer—call it n. Clearly, $n \leq 10^M$. Thus

$$\frac{1}{2^a 5^b} = \frac{n}{10^M},$$

so the decimal expansion of $1/2^a 5^b$ consists of the digits of n, perhaps preceded by some zeros.

* **Exercise 3.** Calculate M for 16, 20, and 25, and compare with the table to see that it gives the correct length of the expansion.

* **Exercise 4.** How many places are there in the expansions of 1/128, 1/320, and 1/800?

The converse of Theorem 1 is also true.

Theorem 2. If $1/n$ has a terminating decimal expansion, then $n = 2^a 5^b$ for some nonnegative integers a and b.

Proof. Let the terminating decimal expansion of $1/n$ be

$$1/n = .d_1 d_2 \cdots d_k$$
$$= d_1/10 + d_2/10^2 + \cdots + d_k/10^k.$$

Then

$$1/n = (d_1 10^{k-1} + d_2 10^{k-2} + \cdots + d_k)/10^k.$$

Call the integer in parentheses m. Then the last equation is

$$1/n = m/10^k \qquad \text{or} \qquad mn = 10^k.$$

The only prime divisors of 10^k are 2 and 5, and so the only prime divisors of n are 2 and 5. This proves the theorem.

Theorems 1 and 2 completely take care of terminating decimals. Among the expansions in the table that do not terminate are some with long periods, including $n = 17$, 19, 23, and 29. In each case, n is a prime and the period of $1/n$ is $n - 1$. But not all primes p have period $p - 1$ for their reciprocals: $1/13$ has period 6, not 12, and $1/11$ has period 2, not 10. As a first step in investigating the lengths of the periods of reciprocals, we prove

Theorem 3. The length of the decimal period of $1/n$ is no longer than $n - 1$.

Proof. Let t be such that $10^t < n < 10^{t+1}$. Then using the division algorithm repeatedly, we have

$$
\begin{aligned}
10^{t+1} &= d_1 n + r_1, & 0 &< r_1 < n, \\
10 r_1 &= d_2 n + r_2, & 0 &\leq r_2 < n, \\
10 r_2 &= d_3 n + r_3, & 0 &\leq r_3 < n, \\
&\quad \cdot & &\quad \cdot \\
&\quad \cdot & &\quad \cdot \\
&\quad \cdot & &\quad \cdot \\
10 r_k &= d_{k+1} n + r_{k+1}, & 0 &\leq r_{k+1} < n, \\
&\quad \cdot & &\quad \cdot \\
&\quad \cdot & &\quad \cdot \\
&\quad \cdot & &\quad \cdot
\end{aligned}
$$

(1)

Note that each d_k is less than 10, because for $k = 2, 3, \ldots,$

$$d_k n = 10 r_{k-1} - r_k \leq 10 r_{k-1} < 10n,$$

and

$$d_1 n = 10^{t+1} - r_1 < 10^{t+1} = 10 \cdot 10^t < 10n.$$

If we divide both sides of the last equation in (1) by $10n$, we get

(2) $r_k/n = d_{k+1}/10 + r_{k+1}/10n$.

If we divide both sides of the first equation in (1) by $10^{t+1}n$ and apply (2) over and over, we get

$$1/n = d_1/10^{t+1} + r_1/n\,10^{t+1}$$
$$= d_1/10^{t+1} + d_2/10^{t+2} + r_2/n\,10^{t+2}$$
$$= d_1/10^{t+1} + d_2/10^{t+2} + d_3/10^{t+3} + r_3/n\,10^{t+3}$$
$$\cdots$$
$$= d_1/10^{t+1} + d_2/10^{t+2} + d_3/10^{t+3} + d_4/10^{t+4} + \cdots,$$

and thus d_1, d_2, d_3, \ldots are the digits in the decimal expansion of $1/n$. For example, for $n = 7$ we have

$$10 = 1 \cdot 7 + 3,$$
$$30 = 4 \cdot 7 + 2,$$
$$20 = 2 \cdot 7 + 6,$$
$$60 = 8 \cdot 7 + 4,$$
$$40 = 5 \cdot 7 + 5,$$
$$50 = 7 \cdot 7 + 1,$$
$$10 = 1 \cdot 7 + 3,$$
$$\cdots,$$

and so the decimal expansion of $1/7$ is $.\overline{142857}$.

Each of the remainders r_1, r_2, \ldots is one of the n values 0, 1, 2, \ldots, $n - 1$. Hence, among the $n + 1$ integers $r_1, r_2, \ldots, r_{n+1}$, there must be two that are equal. If you put $n + 1$ objects in n boxes, one box will contain two objects. If $r_j = r_k$, then it follows from (1) that $d_{k+1} = d_{j+1}$, $d_{k+2} = d_{j+2}$, \ldots, and the decimal repeats, with period no greater than n.

* **Exercise 5.** Apply the division algorithm to find the decimal expansion of $1/41$.

If n is relatively prime to 10, we can get more information about the period of $1/n$.

Theorem 4. If $(n, 10) = 1$, then the period of $1/n$ is r, where r is the smallest positive integer such that $10^r \equiv 1 \pmod{n}$.

Proof. We first note that the integer r exists. The least residues

(mod n) of 1, 10, 10^2, 10^3, . . . , 10^{n-1} may assume only the values 1, 2, 3, . . . , $n - 1$, because no power of 10 is divisible by n. Now we have n objects to be put in $n - 1$ boxes, so one box contains two objects: there exist nonnegative integers a and b, $a \neq b$ and both smaller than n, such that $10^a \equiv 10^b$ (mod n). Dividing the congruence by the smaller power of 10, which is possible since $(n, 10) = 1$, we get r.

Since $10^r \equiv 1$ (mod n), we know that

(3) $$10^r - 1 = kn$$

for some integer k. Written in the base 10, k has at most r digits (because $k < 10^r$). Let

$$k = d_{r-1}d_{r-2} \cdots d_1 d_0 = d_{r-1}10^{r-1} + d_{r-2}10^{r-2} + \cdots + d_1 10 + d_0,$$

where $0 \le d_k < 10$ for $k = 0, 1, . . . , r$. Then from (3)

$$\frac{1}{n} = \frac{k}{10^r - 1} = \frac{d_{r-1}d_{r-2} \cdots d_0}{10^r} \cdot \frac{1}{1 - 10^{-r}}$$

$$= (.d_{r-1}d_{r-2} \cdots d_0)(1 + 10^{-r} + 10^{-2r} + \cdots)$$

$$= .\overline{d_{r-1}d_{r-2} \cdots d_0}.$$

This shows that the period of $1/n$ is *at most* r. (The sequence

$$123123123123. . .$$

repeats after every ninth digit, but its period is three.) We must show that the period is no smaller than r. We can do this by using the above argument backward. Suppose that the period of $1/n$ is s: that is,

$$\frac{1}{n} = .\overline{e_{s-1}e_{s-2} \cdots e_0}$$

for some integers $e_0, e_1, . . . , e_{s-1}$. Then

$$\frac{1}{n} = (.e_{s-1}e_{s-2} \cdots e_0)(1 + 10^{-s} + 10^{-2s} + \cdots)$$

$$= \frac{e_{s-1}e_{s-2} \cdots e_0}{10^s} \cdot \frac{1}{1 - 10^{-s}}$$

$$= \frac{e_{s-1}e_{s-2} \cdots e_0}{10^s - 1} .$$

If we let the integer in the numerator be k', we have $nk' = 10^s - 1$, so $10^s \equiv 1$ (mod n). Since r was the smallest positive integer such that $10^r \equiv 1$ (mod n), this shows that $s \ge r$.

For an example, if n is 21, we have

$$\begin{array}{c|cccccc} n & 1 & 2 & 3 & 4 & 5 & 6 \\ \hline 10^n \ (\mathrm{mod}\ 21) & 10 & 16 & 13 & 4 & 19 & 1 \end{array}$$

so $10^6 = 999999$ is divisible by 21, and in fact, $999999 = 21 \cdot 47619$. Thus

$$\frac{1}{21} = \frac{47619}{999999} = \frac{47619}{1000000}\left(1 - \frac{1}{1 - .000001}\right)$$

$$= (.047619)(1 + (.000001) + (.000001)^2 + \cdots)$$

$$= .047619047619047619 \ldots .$$

* **Exercise 6.** Apply Theorem 4 to find the period of 1/41.

Unfortunately, there are no general rules for looking at an integer and discovering what the number r is. Even among the primes, those for which $r = p - 1$ (7, 17, 19, 23, 29, . . .) are scattered in a pattern no one has yet been able to decipher.

So far we have considered fractions with numerator 1 only. But the general rational number is no more difficult, and the proof of Theorem 4 carries through for fractions c/n, where $(c, n) = 1$. Also, if a fraction is divided by 2 or 5, its period is not changed. Summarizing, we have

Theorem 5. If $n \neq 2^a 5^b$ and $(c, n) = 1$, then the period of the decimal expansion of c/n is r, the smallest positive integer such that

$$10^r \equiv 1 \ (\mathrm{mod}\ n_1),$$

where

$$n = 2^a 5^b n_1$$

and $(n_1, 10) = 1$.

Problems

* 1. Find the periods of the expansions of 1/66, 1/4608, and 1/101.
 2. Find the periods of the expansions of 1/666, 1/925, and 1/1001.
* 3. Find the smallest positive integer r such that $10^r \equiv 1 \ (\mathrm{mod}\ n)$ if n is 33 or 37.
 4. Find the smallest positive integer r such that $10^r \equiv 1 \ (\mathrm{mod}\ n)$ if n is 42 or 45.
 5. Prove that if the decimal expansion of $1/n$ in the base b, that is,

$$1/n = d_1/b + d_2/b^2 + d_3/b^3 + \cdots, \qquad 0 \le d_k < b,$$

terminates, then every prime divisor of n is a divisor of b.

6. Prove that if every prime divisor of n is a divisor of b, then the decimal expansion of $1/n$ in the base b terminates.

 * 7. Which of the reciprocals of 13, 14, . . . , 25 have terminating decimal expansions in the base 12?

8. Prove that if $(n, b) = 1$, then the period of the decimal expansion of $1/n$ in the base b is the smallest positive integer such that $b^r \equiv 1 \pmod{n}$.

 * 9. In the base 2, what is the period of the decimal expansion of (a) 1/3, (b) 1/5, (c) 1/7, (d) 1/9, (e) 1/11?

10. Calculate the decimal expansions in the base 2 of (a) 1/3, (b) 1/5, (c) 1/9.

 * 11. In the base 12, what is the period of the decimal expansion of (a) 1/7, (b) 1/11, (c) 1/17?

12. Calculate the decimal expansion in the base 12 of (a) 1/13, (b) 1/14.

 *† 13. Calculate the following decimal expansions:
(a) $1/9^2$ in the base 10.
(b) $1/7^2$ in the base 8.
(c) $1/6^2$ in the base 7.
(d) Guess a theorem.
(e) Prove it.

14. Show that

$$\sum_{n=1}^{\infty} 7^{-n(n+1)/2}$$

is irrational.

 † 15. Let $(.a_1 a_2 a_3 \ . \ . \ .)_!$ be defined to be

$$a_1/1! + a_2/2! + a_3/3! + \cdots,$$

with $0 \le a_i \le (i - 1)!$ for each i. Does every rational number have a terminating ! expansion? Is it unique?

Section

16

Pythagorean Triangles

More than 3500 years ago the Babylonians knew that the triangle whose sides have length 120, 119, and 169 is a right triangle. They knew many other such triangles too, including those with sides

$$4800, 4601, 6649,$$
$$360, \quad 319, \quad 481,$$
$$6480, 4961, 8161,$$
$$2400, 1679, 2929,$$
$$2700, 1771, 3229;$$

they probably used them as a sort of table of trigonometric functions. They had observed, experimentally, that if x and y denote the lengths of the legs of a right triangle and z denotes the length of its hypotenuse, then $x^2 + y^2 = z^2$. However, it was not until the time of Pythagoras, some 2500 years ago, that it was *proved* that $x^2 + y^2$ always equals z^2, and the result changed from a mere observation like the Law of Gravity (which may be repealed tomorrow for all we know) to an eternal theorem that will never need revision and will never be shown to be false. There will come a day when the works of Shakespeare will be completely forgotten, but the Pythagorean Theorem will still be true: the works of humankind disappear, but mathematical truth endures! It is another reason for studying number theory.

Many proofs of the Pythagorean Theorem are known, and Loomis [8] has compiled more than 400 of them. An Indian proof consisted of the diagram below with the one word: Behold! It is indeed possible to behold that the area of the big square is $(x + y)^2 = x^2 + 2xy + y^2$, and it is also the sum of the areas of the small square and the four triangles, $z^2 + 4(1/2xy) = z^2 + 2xy$, so it follows that $x^2 + y^2 = z^2$.

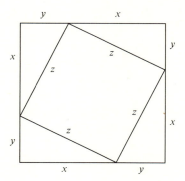

James Garfield constructed a proof before he was President of the United States; mathematical knowledge is thus not a disqualification for high office. A right triangle whose sides are integers we will call a **Pythagorean triangle.** (Strictly speaking, the sides are not integers; they are line segments whose lengths are denoted by integers, but no confusion should arise.) The problem of finding all Pythagorean triangles is the same as that of finding all solutions in positive integers of

(1) $x^2 + y^2 = z^2.$

The ancient Babylonians could find some of the solutions; we will determine them all. There is an immense literature on Pythagorean triangles, and Schaaf [15] and Beiler [2] have references to some of it.

Note first that we may assume that x and y are relatively prime. Suppose not: let $x^2 + y^2 = z^2$ and $(x, y) = d$. Then $d | z$, and

$$(x/d)^2 + (y/d)^2 = (z/d)^2,$$

and we also know that $(x/d, y/d) = 1$. This shows that any solution of (1) may be derived from a solution in which the terms on the left are relatively prime, by multiplication by a suitable factor. Thus when we find all solutions of $x^2 + y^2 = z^2$ with $(x, y) = 1$, we will be able to find all solutions of $x^2 + y^2 = z^2$.

* **Exercise 1.** If $(x, y) = 1$ and $x^2 + y^2 = z^2$, show that $(y, z) = (x, z) = 1$.

We will call a solution $x = a$, $y = b$, $z = c$ of $x^2 + y^2 = z^2$ in which a, b, and c are positive and $(a, b) = 1$ a **fundamental solution.** From Exercise 1 it follows that if a, b, c is a fundamental solution, then no two of a, b, c, have a common prime factor.

Lemma 1. If a, b, c is a fundamental solution of $x^2 + y^2 = z^2$, then exactly one of a and b is even.

Proof. The integers a and b cannot both be even in a fundamental solution.

* **Exercise 2.** Why not?

Nor can a and b both be odd. Suppose that they were. Then $a^2 \equiv 1 \pmod 4$ and $b^2 \equiv 1 \pmod 4$. Thus

$$c^2 = a^2 + b^2 \equiv 2 \pmod 4,$$

which is impossible. The only possibility left is that one of a and b is even and the other is odd.

Corollary. If a, b, c is a fundamental solution, then c is odd.

Proof. $a^2 + b^2 \equiv 1 \pmod 2$.

Before we proceed to derive an expression for all the fundamental solutions of (1), we need to prove

Lemma 2. If $r^2 = st$ and $(s, t) = 1$, then both s and t are squares.

Proof. Write out the prime-power decompositions of s and t:

$$s = p_1^{e_1} p_2^{e_2} \cdots p_k^{e_k}$$
$$t = q_1^{f_1} q_2^{f_2} \cdots q_j^{f_j}.$$

From $(s, t) = 1$, it follows that no prime appears in both decompositions. Because of the unique factorization theorem, the prime-power decomposition of r^2 can be written

$$r^2 = st = p_1{}^{e_1}p_2{}^{e_2} \cdots p_k{}^{e_k}q_1{}^{f_1}q_2{}^{f_2} \cdots q_j{}^{f_j},$$

and the p's and q's are distinct primes. Since r^2 is a square, all of the exponents $e_1, e_2, \ldots, e_k, f_1, f_2, \ldots, f_j$ are even. Thus s and t are squares.

Exercise 3 (optional). Prove Lemma 2 by induction on r as follows. The lemma is trivially true for $r = 1$ and $r = 2$. Suppose that it is true for $r \leq n - 1$. Note that n has a prime divisor p, and $p \,|\, s$ or $p \,|\, t$, but not both. Also, $p^2 \,|\, n^2$. Apply the induction assumption to n/p.

Note that it is impossible to conclude from $r^2 = st$ that s and t are squares if s and t are not relatively prime. For example, $6^2 = 18 \cdot 2$, but neither 18 nor 2 is a square.

The next lemma gives a condition that fundamental solutions of (1) must satisfy.

Lemma 3. Suppose that a, b, c is a fundamental solution of $x^2 + y^2 = z^2$, and suppose that a is even. Then there are positive integers m and n with $m > n$, $(m, n) = 1$ and $m \not\equiv n$ (mod 2) such that

$$a = 2mn,$$
$$b = m^2 - n^2,$$
$$c = m^2 + n^2.$$

(Note that we lose no generality in assuming that a is even. Lemma 1 tells us that exactly one of a and b is even, so we may as well call the even member of the pair a, b by the name of a.)

Proof. Since a is even, $a = 2r$ for some r. So, $a^2 = 4r^2$; from $a^2 = c^2 - b^2$ follows

(2) $$4r^2 = (c + b)(c - b).$$

We know that b is odd, and from the Corollary to Lemma 1, we know that c is odd too. Thus $c + b$ and $c - b$ are both even. Thus we can put

(3) $$c + b = 2s \quad \text{and} \quad c - b = 2t.$$

Then

(4) $$c = s + t \quad \text{and} \quad b = s - t.$$

Substituting (3) into (2), we get $4r^2 = (2s)(2t)$, or

$$r^2 = st.$$

If s and t are relatively prime, we can apply Lemma 2 and conclude that s and t are both squares. In fact, s and t are relatively prime, as we now show. Suppose that $d \mid s$ and $d \mid t$. From (4) it follows that $d \mid b$ and $d \mid c$. But from Exercise 1, we know that b and c are relatively prime. Hence $d = \pm 1$ and $(s, t) = 1$. Lemma 2 says that

$$s = m^2 \quad \text{and} \quad t = n^2$$

for some integers m and n, which we may assume to be positive. Thus

$$a^2 = 4r^2 = 4st = 4m^2n^2$$

or $a = 2mn$. From (4),

$$c = s + t = m^2 + n^2,$$
$$b = s - t = m^2 - n^2.$$

Having established the last three equations, we need now only establish that $m > n$, $(m, n) = 1$, and $m \not\equiv n \pmod 2$ to complete the proof. The inequality follows because b is part of a fundamental solution and hence positive.

Exercise 4. Suppose that $d \mid m$ and $d \mid n$. Show that $d \mid a$ and $d \mid b$. Conclude that $(m, n) = 1$.

Exercise 5. Suppose that $m \equiv n \equiv 0 \pmod 2$. Show that a and b are both even, which is impossible.

Exercise 6. Suppose that $m \equiv n \equiv 1 \pmod 2$. Show again that a and b are both even.

Exercises 4, 5, and 6 have completed the proof of Lemma 3. For example, 33, 56, 65 is a fundamental solution, since $1089 + 3136 = 4225$, and from $r^2 = (56/2)^2 = 28^2 = 7^2 \cdot 4^2$ we get $m = 7$ and $n = 4$. They are the right values, since $m^2 - n^2 = 49 - 16 = 33$ and $m^2 + n^2 = 49 + 16 = 65$.

We have shown that if a, b, c is a fundamental solution, than a, b, and c satisfy the conditions of Lemma 3. It is also true that if a, b, c satisfy the conditions of Lemma 3, then a, b, c is a fundamental solution of $x^2 + y^2 = z^2$. We prove this in

Lemma 4. If

$$a = 2mn,$$
$$b = m^2 - n^2,$$
$$c = m^2 + n^2,$$

then a, b, c is a solution of $x^2 + y^2 = z^2$. If in addition, $m > n$, m and n are positive, $(m, n) = 1$, and $m \not\equiv n$ (mod 2), then a, b, c is a fundamental solution.

Proof. To see that a, b, c is a solution is a matter of computation:

$$a^2 + b^2 = (2mn)^2 + (m^2 - n^2)^2$$
$$= 4m^2n^2 + m^4 - 2m^2n^2 + n^4 = m^4 + 2m^2n^2 + n^4$$
$$= (m^2 + n^2)^2 = c^2.$$

It remains to show that $(m, n) = 1$ and $m \not\equiv n$ (mod 2) imply that $(a, b) = 1$. Suppose that p is an odd prime such that $p|a$ and $p|b$. From $c^2 = a^2 + b^2$ it follows that $p|c$. From $p|b$ and $p|c$ it follows that $p|(b + c)$ and $p|(b - c)$. But

$$b + c = 2m^2 \quad \text{and} \quad b - c = -2n^2.$$

So, $p|2m^2$ and $p|2n^2$. Since p is odd, this implies that $p|m^2$ and $p|n^2$, and hence that $p|m$ and $p|n$. Since m and n are relatively prime, this is impossible. The only way in which a and b could fail to be relatively prime is for both to be divisible by 2. But b is odd because $b = m^2 - n^2$, and one of m, n is even and the other is odd. Thus $(a, b) = 1$, and because $m > n$, b is positive. Because m and n are positive, a is positive. Thus a, b, c is a fundamental solution.

We restate Lemmas 3 and 4 as

Theorem 1. All solutions $x = a$, $y = b$, $z = c$ to $x^2 + y^2 = z^2$, where a, b, c are positive and have no common factor and a is even, are

given by

$$a = 2mn,$$
$$b = m^2 - n^2,$$
$$c = m^2 + n^2,$$

where m and n are any relatively prime integers, not both odd, and $m > n$.

Here is a table of some fundamental Pythagorean triangles with small sides:

m	n	a	b	c	a^2	b^2	c^2
2	1	4	3	5	16	9	25
3	2	12	5	13	144	25	169
4	1	8	15	17	64	225	289
4	3	24	7	25	576	49	625
5	2	20	21	29	400	441	841
5	4	40	9	41	1600	81	1681
6	1	12	35	37	144	1225	1369
7	2	28	45	53	784	2025	2809

Problems

* 1. There are eleven nonfundamental Pythagorean triangles with hypotenuses less than 50; find them.

2. Find a fundamental Pythagorean triangle with hypotenuse 265.

* 3. Find a fundamental Pythagorean triangle with leg 100.

4. How many Pythagorean triangles (fundamental or not) can you find with hypotenuse 1105?

5. If $(a, b) = d$ and $a^2 + b^2 = c^2$, show that $(a, c) = (b, c) = d$.

6. If $(a, b) = 1$ and $ab = c^n$, show that a and b are nth powers.

7. If $(a, b) = d$ and $ab = c^n$, show that a/d and b/d are not necessarily nth powers.

8. Bhascara (ca. 1150) found a right triangle whose area is numerically equal to the length of its hypotenuse. Show that this cannot happen if the triangle has integer sides.

* 9. In all of the Pythagorean triangles in the table in the text, one side is a multiple of five. Is this true for all Pythagorean triangles?

10. Show that 12 divides the product of the legs of a Pythagorean triangle.

11. Show that 60 divides the product of the sides of a Pythagorean triangle.

12. Here is a quadrilateral, not a parallelogram, with integer sides and integer area:

(a) What is its area?
(b) Such quadrilaterals are not common; can you find another?
(c) Could you find 1,000,000 more?

*13. Find all fundamental Pythagorean triangles whose areas are twice their perimeters.

14. Find all fundamental Pythagorean triangles whose areas are three times their perimeters.

† 15. Prove that 3, 4, 5 is the only solution of $x^2 + y^2 = z^2$ in consecutive positive integers.

16. Show that the only Pythagorean triangles with sides in arithmetic progression are those with sides $3n$, $4n$, $5n$, $n = 1, 2, 3, \ldots$.

† 17. $3^2 + 4^2 = 5^2$. $5^2 + 12^2 = 13^2$. $7^2 + 24^2 = 25^2$. $9^2 + 40^2 = 41^2$.

(a) Guess a theorem.
(b) Prove that the numbers in (a) give the *only* Pythagorean triangles with consecutive integers for one leg and a hypotenuse.

18. (a) Look in the table in the text and find two Pythagorean triangles with the same area.
(b) Can you find two others with the same area?
(c) Prove that two Pythagorean triangles with the same area and equal hypotenuses are congruent.

19. Show that $n^2 + (n + 1)^2 = 2m^2$ is impossible.

20. (a) $3^2 + 4^2 = 5^2$. $20^2 + 21^2 = 29^2$. $119^2 + 120^2 = 169^2$. To find another such relation, show that if $a^2 + (a + 1)^2 = c^2$, then

$$(3a + 2c + 1)^2 + (3a + 2c + 2)^2 = (4a + 3c + 2)^2.$$

(b) If $a^2 + (a + 1)^2 = c^2$, let $u = c - a - 1$ and $v = (2a + 1 - c)/2$. Show that v is an integer and that $u(u + 1)/2 = v^2$. This shows that there are infinitely many square triangular numbers.

17

Infinite Descent and Fermat's Conjecture

In the section on Pythagorean triangles, we found all of the solutions in integers of $x^2 + y^2 = z^2$. After disposing of that problem, it would be natural to try the same ideas on an equation of one higher degree, $x^3 + y^3 = z^3$. The same ideas would not work, nor would any others; there is no solution in integers of $x^3 + y^3 = z^3$. There is one exception—a solution in which one of the variables is zero. We will call such a solution a **trivial** solution, treating it with the contempt it deserves. When we say "solution" in this section, we will mean "nontrivial solution." In fact, no one knows any solution in integers to any of the equations $x^n + y^n = z^n$ for $n \geq 3$. Fermat thought he had a proof that $x^n + y^n = z^n$ has no nontrivial solutions when $n \geq 3$; he wrote a note in the margin of his copy of the works of Diophantus saying that he had a proof, but that the margin was too small to contain it. It is almost certain that he was mistaken, but of course we cannot be certain. He may have had a proof. Or, he may have realized how deep the proof must lie and wrote the comment to keep future generations of mathematicians at work. Could he have been such a practical joker? It is not likely: Fermat was a judge, and judges tend to be sober and not given to pranks.

The statement "$x^n + y^n = z^n$ has no nontrivial solutions if $n \geq 3$" is

often called "Fermat's Last Theorem"—to distinguish it from the theorem that bears his name (see Section 6), but a better name would be *Fermat's Conjecture*. An enormous amount of work has been devoted to it, but it is still not settled one way or the other. The conjecture is known to be true for $n < 25,000$ and for many larger values of n, but this is far from a proof. Before the First World War, there was a large prize offered in Germany for a correct proof, and many amateurs offered attempted solutions. It is said that the great number theorist Landau had postcards printed which read, "Dear Sir or Madam: Your attempted proof of Fermat's Theorem has been received and is herewith returned. The first mistake is on page _____ , line _____." Landau would give them to his students to fill in the missing numbers. Even though there is no longer a prize for the solution, mathematical amateurs still attempt proofs, and many convince themselves that they have succeeded. They then try to convince mathematicians, fail, and sometimes become quite bitter about what they think is a conspiracy of mathematicians to keep them from getting the recognition due them. It is sad. Besides amateurs, many powerful mathematicians have worked on the problem, and it may be that Fermat's Conjecture is forever undecidable one way or the other. There may exist a solution of $x^n + y^n = z^n$ in numbers so large that no one could ever find them. There are, after all, integers so big that the world could not hold them if they were written out. If they take up that much room, can they fit into the head of man?

The object of this section is to show that Fermat's Conjecture is true for $n = 4$, and in so doing, to illustrate Fermat's method of infinite descent. You might wonder why we avoid considering the case $n = 3$. It turns out to be harder to show that $x^3 + y^3 = z^3$ has no solutions than it is to show that $x^4 + y^4 = z^4$ has none, though it is also possible to apply the method of infinite descent when $n = 3$. We will prove

Theorem 1. There are no nontrivial solutions of

$$x^4 + y^4 = z^2.$$

Note that this implies that there are no solutions of $x^4 + y^4 = z^4$; if a, b, c were a solution to that equation, we would have $a^4 + b^4 = (c^2)^2$, contrary to Theorem 1.

Proof. We will apply Fermat's method of infinite descent. Consider the nontrivial solutions of $x^4 + y^4 = z^2$. We want to show that there are none. We will suppose that there is one and deduce a contradiction. Among the nontrivial solutions, there is one with a smallest value of z^2. Let c^2 denote this value of z^2. There may be several solutions with this same value of z; if there are, we will pick any one of them—it makes no difference which. Call the solution that we pick a, b, c. The idea of the proof is to construct numbers r, s, t that also satisfy $x^4 + y^4 = z^2$, with $t^2 < c^2$. Since c^2 was chosen as small as possible, it follows that the assumption that there were nontrivial solutions was wrong. Hence there are no nontrivial solutions. This method is no mere trick, but is quite natural. It is very possible that Fermat one day set himself to the task of finding solutions of $x^4 + y^4 = z^4$; he may have applied various devices to reduce the equation in the same way we reduced the equation $x^2 + y^2 = z^2$; and maybe he was surprised when the result of his efforts was another equation of the same form, but with smaller numbers—surprised and pleased too, because this allowed him to conclude that if there was a solution, then there was another smaller solution, and then another and another and another: an infinitely descending chain of solutions. But since we may assume that $x, y,$ and z are positive, this is impossible. (On the other hand, Fermat may have sat down and thought, "I will now apply my method of infinite descent to $x^4 + y^4 = z^4$"; history is silent on the subject.)

We suppose that we have a nontrivial solution a, b, c, with c^2 as small as possible. We can suppose that a and b are relatively prime. Suppose not. Then there is a prime p such that $p \mid a$ and $p \mid b$, and hence $p^2 \mid c$. Thus $(a/p)^4 + (b/p)^4 = (c/p^2)^2$ provides a solution to $x^4 + y^4 = z^2$ with a smaller value of z^2 than c^2, and we have supposed this to be impossible.

* **Exercise 1.** Show that a and b cannot both be odd. (Consider $a^4 + b^4 = c^2$ modulo 4.)

Because $(a, b) = 1$, a and b cannot both be even, either. Thus one is even, and look at $b^2 = m^2 - n^2$ modulo 4. Remember that $x^2 \equiv -1$ agree to call the even member of the pair a, b by the name of a. But now we have a fundamental solution of $x^2 + y^2 = z^2$ as defined in Section 16:

$$(a^2)^2 + (b^2)^2 = c^2,$$

$(a^2, b^2) = 1$, and a^2 is even and b^2 is odd. Hence, by Lemma 3 of Section 16, there are integers m and n, relatively prime and not both odd, such that

$$a^2 = 2mn,$$
(1)
$$b^2 = m^2 - n^2,$$
$$c = m^2 + n^2.$$

Exercise 2. Show that n is even. (Suppose that n is odd and m is even, and look at $b^2 = m^2 - n^2$ modulo 4. Remember that $x^2 \equiv -1$ (mod 4) is impossible.)

Because n is even, it follows that m is odd. Put

$$n = 2q.$$

Then from (1), $a^2 = 4mq$, or

(2)
$$\left(\frac{a}{2}\right)^2 = mq.$$

We would like to conclude that m and q are both squares. To do that, we need, according to Lemma 2 of the last section, to show that m and q are relatively prime.

Exercise 3. Show that $(m, q) = 1$. (Suppose not, and deduce that $(m, n) \neq 1$.)

So, there are integers t and v such that

$$m = t^2 \qquad \text{and} \qquad q = v^2.$$

Exercise 4. Verify that t and v are relatively prime. (Suppose not, and deduce that $(m, q) \neq 1$.)

Exercise 5. Note that t is odd. (Suppose not, and deduce that m is even.)

So far, we have found out a good many things about a, b, and c. We need more yet. We start with the obvious observation

$$n^2 + (m^2 - n^2) = m^2.$$

Substituting into this the various facts we know, namely

$$n = 2q = 2v^2, \qquad m^2 - n^2 = b^2, \qquad m = t^2,$$

the equation becomes

$$(2v^2)^2 + b^2 = (t^2)^2,$$

so we have another Pythagorean triangle. Is $2v^2$, b, t^2 a fundamental solution? It is if $2v^2$ and b are relatively prime, and they are.

* **Exercise 6.** Supply the reasons for the following implications:

$$\text{if } p \mid 2v^2 \text{ and } p \mid b, \text{ then } p \mid n \text{ and } p \mid b,$$
$$\text{if } p \mid n \text{ and } p \mid b, \text{ then } p \mid n \text{ and } p \mid m,$$
$$\text{if } (m, n) = 1, \text{ then } (2v^2, b) = 1.$$

With $2v^2$ even, we have a fundamental solution of $x^2 + y^2 = z^2$, so we can apply Lemma 3 of the last section to show that there are integers M and N, with $(M, N) = 1$ and $M \not\equiv N \pmod 2$, such that

$$2v^2 = 2MN,$$
(3) $$b = M^2 - N^2,$$
$$t^2 = M^2 + N^2.$$

Thus we have $v^2 = MN$ and $(M, N) = 1$. The product of two relatively prime integers is a square if and only if both integers are squares (Lemma 2 of the last section), so there are integers r and s such that

$$M = r^2 \qquad \text{and} \qquad N = s^2.$$

From the third equation in (3), we have

$$t^2 = (r^2)^2 + (s^2)^2,$$

or

$$r^4 + s^4 = t^2.$$

Here is another solution of $x^4 + y^4 = z^2$. It has the property that

$$t^2 = m \leq m^2 < m^2 + n^2 = c \leq c^2,$$

which is impossible, because c^2 was chosen as small as possible. This contradiction proves the theorem.

* **Exercise 7.** Why $m^2 < m^2 + n^2$? Why not just $m^2 \leq m^2 + n^2$?

Here is another example of the method of infinite descent. Suppose that there are positive integers a and b such that

(4) $$\sqrt{3} = \frac{a}{b},$$

where the fraction is in lowest terms, so that $(a, b) = 1$. From (4), $a^2 = 3b^2$, so

$$3(a - b)^2 = 3a^2 - 6ab + 3b^2 = 9b^2 - 6ab + a^2 = (3b - a)^2,$$

from which we get

(5)
$$\sqrt{3} = \frac{3b - a}{a - b}.$$

Since $1 < \sqrt{3} < 2$, from (4) we have $b < a < 2b$ or $0 < a - b < b$. Thus the denominator in (5) is smaller than the denominator in (4), so we can start with a fraction equal to $\sqrt{3}$ and descend through infinitely many others with smaller and smaller denominators. Since this is impossible, we conclude that $\sqrt{3}$ is not equal to a quotient of two integers, and so is irrational.

This process can be reversed to ascend to better and better rational approximations to $\sqrt{3}$. If we start with 3/2 and put $(3b - a)/(a - b) = 3/2$, we get $a/b = 9/5$. Continuing, we get the sequence 3/2, 9/5, 12/7, 33/19, 45/26, . . . or 1.5, 1.8, 1.714, 1.737, 1.731, . . . , converging to $\sqrt{3} = 1.732050808.$. . .

Problems

1. Use the method of the text to show that \sqrt{n} is irrational if n is not a perfect square.

2. Why does the method of the text fail to show that \sqrt{n} is irrational if n is a perfect square?

*† 3. Does $x^8 + y^8 = z^8$ have nontrivial solutions?

4. Use the method of infinite descent to show that $x^3 + 3y^3 = 9z^3$ has no nontrivial solutions.

* 5. Generalize: does $x^3 + py^3 = p^2z^3$ have nontrivial solutions?

6. Generalize: does $x^4 + py^4 + p^2z^4 = p^3w^4$ we have nontrivial solutions?

* 7. Generalize: does $x_1^n + px_2^n + p^2x_3^n + \cdots + p^{n-2}x_{n-1}^n = p^{n-1}x_n^n$ have nontrivial solutions?

8. Show that $x^2 + y^2 + z^2 = 2xyz$ has no nontrivial solutions.

* 9. For what values of k does the method used in solving Problem 8 show that there are no nontrivial solutions to $x^2 + y^2 + z^2 = kxyz$?

10. Show that there are no nontrivial solutions to $x^2 + y^2 = x^2y^2$ or $x^2 + y^2 + z^2 = x^2y^2$.

Sums of Two Squares

Among the integers from 1 to 99, the following 57 are not representable as the sum of two squares of integers:

3	6	7	11	12	14	15	19	21	22	23
24	27	28	30	31	33	35	38	39	42	43
44	46	47	48	51	54	55	56	57	59	60
62	63	66	67	69	70	71	75	76	77	78
79	83	84	86	87	88	91	92	93	94	95
96	99									

But the remaining 42 are:

1	2	4	5	8	9	10	13	16	17	18
20	25	26	29	32	34	36	37	40	41	45
49	50	52	53	58	61	64	65	68	72	73
74	80	81	82	85	89	90	97	98		

It would be an exercise of your inductive powers to look at these lists and, in the spirit of the scientific method, try to formulate a hypothesis that would explain the presence of a number on its list and which could be used to predict results for numbers greater than 99. There is a fairly

simple property (other than not being representable as a sum of two squares) that the numbers in the first list share. It is not to your discredit if you cannot observe what it is, since probably only one mind in a million would have the power and training necessary to see it; but once one mind sees it, others can understand it, see why it is so, appreciate it, and apply it. It is

Theorem 1. n cannot be written as the sum of two squares if and only if the prime-power decomposition of n contains a prime congruent to 3 (mod 4) to an odd power.

Proof (of the "if" part). Suppose that p is a prime, $p \equiv 3$ (mod 4), which appears in the prime-power decomposition of n to an odd power. That is, for some $e \geq 0$, $p^{2e+1} | n$ and $p^{2e+2} \nmid n$. Suppose that $n = x^2 + y^2$ for some x and y. We will deduce a contradiction—namely that -1 is a quadratic residue (mod p). Let $d = (x, y)$, $x_1 = x/d$, $y_1 = y/d$, and $n_1 = n/d^2$. Then

$$(1) \qquad x_1{}^2 + y_1{}^2 = n_1 \qquad \text{and} \qquad (x_1, y_1) = 1.$$

If p^f is the highest power of p that divides d, then n_1 is divisible by $p^{2e-2f+1}$, and this exponent, being odd and nonnegative, is at least one. Thus $p | n_1$, and if $p | x_1$, it follows from (1) that $p | y_1$. But $(x_1, y_1) = 1$, so $p \nmid x_1$. Hence there is a number u such that

$$x_1 u \equiv y_1 \text{ (mod } p).$$

Thus

$$(2) \qquad 0 \equiv n_1 \equiv x_1{}^2 + y_1{}^2 \equiv x_1{}^2 + (ux_1)^2 \equiv x_1{}^2(1 + u^2) (\text{mod } p).$$

Since $(x_1, p) = 1$, x_1 may be cancelled in (2) to give

$$1 + u^2 \equiv 0 \text{ (mod } p).$$

This says that -1 is a quadratic residue (mod p), which is impossible, since $p \equiv 3$ (mod 4). Hence $n = x^2 + y^2$ is impossible, and the easy part of the theorem is proved.

The rest of Theorem 1 (the "only if" part) is harder. We will need four lemmas.

Lemma 1. $(a^2 + b^2)(c^2 + d^2) = (ac + bd)^2 + (ad - bc)^2$ for any integers $a, b, c,$ and d.

Proof. Multiply it out.

The result shows that if two numbers are representable as sums of two squares, then so is their product. Rather than "n is representable as the sum of two squares of integers," we will say "n is representable" for short.

Lemma 2. If n is representable, then so is k^2n for any k.

Proof. If $n = x^2 + y^2$, then $k^2n = (kx)^2 + (ky)^2$.

Lemma 3. Any integer n can be written in the form

$$n = k^2 p_1 p_2 \cdots p_r,$$

where k is an integer and the p's are different primes.

Exercise 1. Convince yourself that this is so by considering the prime-power decomposition of n.

As an example of the application of Lemmas 1 and 2, we can get a representation for $260 = 2^2 \cdot 5 \cdot 13$ from the representations

$$5 = 2^2 + 1^2 \qquad \text{and} \qquad 13 = 3^2 + 2^2.$$

From Lemma 1,

$$\begin{aligned} 65 = 5 \cdot 13 &= (2^2 + 1^2)(3^2 + 2^2) \\ &= (2 \cdot 3 + 1 \cdot 2)^2 + (2 \cdot 2 - 1 \cdot 3)^2 = 8^2 + 1^2. \end{aligned}$$

Hence

$$260 = 2^2 \cdot 65 = (8 \cdot 2)^2 + (1 \cdot 2)^2 = 16^2 + 2^2.$$

Exercise 2. Write 325 as a sum of two squares.

Exercise 3. If the prime-power decomposition of n contains no prime p, $p \equiv 3 \pmod 4$, to an odd power, then note that

$$n = k^2 p_1 p_2 \cdots p_r \qquad \text{or} \qquad n = 2k^2 p_1 p_2 \cdots p_r$$

for some k and r, where each p is congruent to 1 (mod 4).

Exercise 3 and Lemmas 1 and 2 show that to prove the rest of Theorem 1, it is sufficient to prove

Lemma 4. Every prime congruent to 1 (mod 4) can be written as a sum of two squares.

Proof. The idea of the proof is this: if $p \equiv 1$ (mod 4), then we show that there are integers x and y such that

$$x^2 + y^2 = kp$$

for some integer k, $k \geq 1$. If $k > 1$, we then construct from x and y new integers x_1 and y_1 such that

$$x_1{}^2 + y_1{}^2 = k_1 p$$

for some k_1, with $k_1 < k$. This is the step that proves the lemma, because if $k_1 > 1$, we repeat the process to get integers x_2 and y_2 such that $x_2{}^2 + y_2{}^2 = k_2 p$ with $k_2 < k_1$. If we keep on, we will get a descending sequence of positive integers, $k > k_1 > k_2 > \cdots$, which cannot be infinite: it must eventually reach 1. When it does, we have a representation of p as a sum of two squares.

First we show that we can find x and y such that $x^2 + y^2 = kp$ for some k, $k \geq 1$. Since $p \equiv 1$ (mod 4), we know that -1 is a quadratic residue (mod p); hence there is an integer u such that $u^2 \equiv -1$ (mod p). That is, $p \mid (u^2 + 1)$, or

$$u^2 + 1 = kp$$

for some k, $k \geq 1$. Hence $x^2 + y^2 = kp$ always has a solution for some k, $k \geq 1$; in fact, we can take $y = 1$. For example, if $p = 17$, we have $4^2 + 1^2 = 1 \cdot 17$; if $p = 29$, $12^2 + 1^2 = 5 \cdot 29$. The number u can be found by trial (we can write $kp - 1$ for $k = 1, 2, \ldots$, and continue until we come to a square), or we can use the fact that

$$\left(\left(\frac{p-1}{2} \right)! \right)^2 \equiv -1 \ (\text{mod } p).$$

The last congruence (which follows from Wilson's Theorem) gives a construction for u: a long one, perhaps, but one which guarantees a result.

We now show how to construct x_1 and y_1. Define r and s by

(3) $r \equiv x$ (mod k), $s \equiv y$ (mod k), $-\dfrac{k}{2} < r \leq \dfrac{k}{2}$, $-\dfrac{k}{2} < s \leq \dfrac{k}{2}$.

From (3),

$$r^2 + s^2 \equiv x^2 + y^2 \ (\text{mod } k).$$

But we had chosen x and y such that $x^2 + y^2 = kp$. Hence

$$r^2 + s^2 \equiv 0 \pmod{k},$$

or

(4) $$r^2 + s^2 = k_1 k$$

for some k_1. It follows from (4) that

$$(r^2 + s^2)(x^2 + y^2) = (k_1 k)(kp) = k_1 k^2 p.$$

From Lemma 1, however, we have

$$(r^2 + s^2)(x^2 + y^2) = (rx + sy)^2 + (ry - sx)^2.$$

Thus

(5) $$(rx + sy)^2 + (ry - sx)^2 = k_1 k^2 p.$$

Note that from (3),

$$rx + sy \equiv r^2 + s^2 \equiv 0 \pmod{k}$$

and

$$ry - sx \equiv rs - sr \equiv 0 \pmod{k}.$$

Thus k^2 divides each term on the left-hand side of (5); dividing (5) by k^2, we get

$$\left(\frac{rx + sy}{k}\right)^2 + \left(\frac{ry - sx}{k}\right)^2 = k_1 p,$$

an equation in integers. Let $x_1 = (rx + sy)/k$ and $y_1 = (ry - sx)/k$. Then $x_1^2 + y_1^2 = k_1 p$, and the lemma will be proved when we show that $k_1 < k$. The inequalities in (3) give

$$r^2 + s^2 \leq (k/2)^2 + (k/2)^2 = k^2/2.$$

But

$$r^2 + s^2 = k_1 k.$$

Thus $k_1 k \leq k^2/2$ or $k_1 \leq k/2$. Hence $k_1 < k$, and to complete the proof of the lemma, we need only to note that $k_1 \geq 1$. If $k_1 = 0$, then from the last equation, $r = s = 0$. It follows from (3) that $k|x$ and $k|y$. Thus $k|p$, so $k = 1$ or p. If $k = 1$, then $x^2 + y^2 = p$ and the lemma is proved, and if $k = p$, then $u^2 + 1 = p^2$, which is impossible.

Let us take an example. Starting with $12^2 + 1^2 = 5 \cdot 29$, we will carry through the calculations of the lemma to get a representation of 29 as a sum of two squares. We have $x = 12, y = 1$, and $k = 5$. We have $r \equiv 12$

(mod 5) and $s \equiv 1$ (mod 5); choosing r and s in the proper range, we get $r = 2$ and $s = 1$. Then

$$5^2 \cdot 29 = (2^2 + 1^2)(12^2 + 1^2) = (2 \cdot 12 + 1 \cdot 1)^2 + (2 \cdot 1 - 1 \cdot 12)^2$$
$$= 25^2 + 10^2.$$

Dividing by $k^2 = 25$ gives $29 = 5^2 + 2^2$, the desired representation.

Exercise 4. Try the calculation for $23^2 + 1^2 = 10 \cdot 53$.

We will end with some remarks on diophantine equations closely related to the sum of two squares. We have completely solved the problem of representing integers as the sum of two squares: we know which integers can be so represented, and Lemma 4, combined with earlier lemmas, gives a method for actually calculating the representation. It is natural now to wonder about the representations of integers as the sum of *three* squares. We would expect that more integers can be represented when we have an extra square to add, and this is the case. It is a fact that n can be written as a sum of three squares, unless $n = 4^e(8k + 7)$ for some integers e and k. So, the numbers smaller than 100 which cannot be written as the sum of three squares are

7, 15, 23, 28, 31, 39, 47, 55, 60, 63, 71, 79, 87, 92, 95;

a total of 15, as against 57 when only two squares were allowed. There are even fewer exceptions if we look at sums of *four* squares; in fact, there are none at all. Every integer can be written as a sum of four squares, as we will show in the next section.

That would seem to settle the squares. That is, as far as the mere representation is concerned. Of course, there are many, many other questions that can be asked, and some that can be answered. For example, how many representations does an integer have as a sum of two squares? What is the sum of the number of solutions of $x^2 + y^2 = n^2$ for $n = 1, 2, \ldots, N$? And so on; as soon as one question is settled, others crowd in to take its place. What about cubes? It is true that every integer can be written as a sum of nine cubes. No one knows what the corresponding number is for the sum of fourth powers (though it is known that every integer greater than $10^{10^{89}}$ is a sum of 19 fourth powers), but the answer is known for fifth, sixth, seventh, and almost all higher powers—for example, 37 fifth powers will do. Let $g(k)$ be the least value of s such that every integer can be written as a sum of no more than s kth powers. The problem of finding $g(k)$ is called *Waring's*

Problem, after a mathematician who wrote in 1770 that every integer was the sum of 4 squares, 9 cubes, 19 fourth powers, and so on. He was just guessing; it was not until 1909 that David Hilbert proved that the number $g(k)$ existed for each k, and even then almost nothing was known about its size for large k. Now $g(k)$ can be determined for all values of k, except for $k = 4$. For all k, $1 \le k \le 200000$, except $k = 4$, $g(k) = 2^k + [(3/2)^k] - 2$, and it is strongly suspected that this formula is true for all $k \ge 1$. (The notation $[(3/2)^k]$ denotes the largest integer less than $(3/2)^k$, so for example $[(3/2)^3] = [27/8] = 3$.)

Not every integer needs nine cubes to represent it: in fact, only 23 and 239 require so many. The largest integer that needs eight cubes is 454, and if you make a table of representations of integers as sums of cubes, you will see, as did Jacobi when he made one by hand in 1851, that the number of integers requiring seven cubes decreases as the size of the integers increase. Numbers like 23, 239, and 454 are only annoying exceptions: of more interest than $g(k)$ is $G(k)$, the least value of s such that every *sufficiently large* integer can be written as a sum of no more than s kth powers. It is known that $G(2) = 4$, and it is curious that the only other value of k for which $G(k)$ is known is 4, the only small value of k for which $g(k)$ is known. $G(4) = 16$, and the most that is known about other k are upper bounds: $G(3) \le 7$ and $G(5) \le 23$, for example. The most that is known in general is that $k + 1 \le G(k) \le k(3 \ln k + 11)$, and the right-hand inequality is extremely difficult and complicated to prove. Work on Waring's Problem continues today.

As it does on another famous problem about representing integers as sums of certain other integers: *Goldbach's Conjecture.* In 1742, an amateur mathematician named Goldbach noted that

$$4 = 2 + 2, \qquad 6 = 3 + 3, \qquad 8 = 5 + 3, \qquad 10 = 5 + 5,$$
$$12 = 7 + 5, \qquad 14 = 7 + 7, \qquad \ldots, \qquad 100 = 97 + 3,$$

and guessed that every even integer greater than two could be written as the sum of two primes. He asked Euler if he could prove it. Euler failed, and no one since has succeeded. The conjecture is almost certainly true—large even integers have *thousands* of representations—and it is infuriating that it cannot be proved. The most powerful tools of number theory have been applied, but they have given only partial results, such as that every sufficiently large even integer is the sum of no more than four primes, or is the sum of a prime and an integer that has no more than two different prime factors, or that every even integer

greater than two is a sum of no more than 20000000000 primes. The course of human history will not be altered much if Goldbach's Conjecture is settled one way or the other, but it would be nice to find out.

Problems

* 1. Determine which of 1980, 1981, 1982, 1983, and 1984 can be written as a sum of two squares, and for those that can, find a representation.

 2. Determine which of 2000, 2001, 2002, 2003, and 2004 can be written as a sum of two squares, and for those that can, find a representation.

* 3. Use Lemma 1 to find all eight of the representations of $32045 = 5 \cdot 13 \cdot 17 \cdot 29$ as sums of two squares.

 4. Use Lemma 1 to get two representations of 4453 as a sum of two squares.

 5. Prove that if $n \equiv 3 \pmod 4$, then n cannot be represented as a sum of two squares.

 6. What can n be congruent to (mod 9) if n is a sum of two squares?

* 7. Is it true that if m and n are sums of two squares and $m \mid n$, then n/m is a sum of two squares? Prove or give a counterexample.

 8. Is it true that if m and n are sums of two squares, then mn is a sum of two squares? Prove or give a counterexample.

 9. Fermat wrote:
 $2n + 1$ is the sum of two squares when and only when (i) n is even, and (ii) $2n + 1$, when divided by the largest square entering into it as a factor must not be divisible by a prime $4k - 1$.
 Show that this is equivalent to Theorem 1.

 10. Girard wrote in 1632 that the numbers representable as the sum of two squares comprise every square, every prime $4k + 1$, a product of such numbers, and the double of any of the preceding. Show that this is equivalent to Theorem 1.

19

Sums of Four Squares

In this section we will prove that every positive integer can be written as a sum of four squares of integers, some of which may be zero. This theorem is quite old. Diophantus seems to have assumed that every positive integer is a sum of 2, 3, or 4 squares of positive integers, but he never explicitly stated the theorem. The first to do so was Bachet (1621). He verified that it was true for integers up to 325, but he was unable to prove it. Fermat said that he was able to prove it using his method of descent; as usual, he gave no details. In the light of subsequent work on the theorem, we may doubt that Fermat's proof was complete. Descartes said that the theorem was no doubt true, but he judged the proof "so difficult that I dared not undertake to find it." (It is hard to resist the temptation to read "so difficult that I was unable to find it.")

Euler next took up the challenge, first working on the problem in 1730. In 1743 he noted that the product of two sums of four squares is again a sum of four squares, a result fundamental to the proof of the theorem, and indeed, he proved the theorem except for one point. In 1751, still pursuing that point, Euler proved another fundamental result, namely that $1 + x^2 + y^2 \equiv 0 \pmod{p}$ always has a solution for any prime p. But the theorem was still out of his reach. Finally, in 1770,

Lagrange, drawing heavily on Euler's ideas, succeeded in constructing a proof. In 1773, Euler (then 66 years old) gave a simpler proof— success after 43 years.

We start by proving Euler's two results.

Lemma 1. The product of two sums of four squares is a sum of four squares.

Proof. The proof is utterly trivial. Finding the result was quite another matter, as witness the thirteen-year gap between Euler's first attack on the problem and his discovery of the following identity:

$$(a^2 + b^2 + c^2 + d^2)(r^2 + s^2 + t^2 + u^2) = (ar + bs + ct + du)^2$$
$$+ (as - br + cu - dt)^2 + (at - bu - cr + ds)^2$$
$$+ (au + bt - cs - dr)^2.$$

It may be verified by multiplication. Note that the right-hand side, when multiplied out, contains all the terms a^2r^2, a^2s^2, . . . , d^2u^2 that appear on the left-hand side multiplied out. Note also that it is not impossible to see by inspection that all the cross-product terms vanish.

To show that every positive integer is the sum of four squares, it follows from Lemma 1 that we need only show that every prime is the sum of four squares. For example, from

$$37 = 6^2 + 1^2 + 0^2 + 0^2 \qquad \text{and} \qquad 57 = 7^2 + 2^2 + 2^2 + 0^2,$$

we get

$$2109 = 57 \cdot 37 = (6 \cdot 7 + 1 \cdot 2 + 0 \cdot 2 + 0 \cdot 0)^2 + (6 \cdot 2 - 1 \cdot 7 + 0 \cdot 0 - 0 \cdot 2)^2$$
$$+ (6 \cdot 2 - 1 \cdot 0 - 0 \cdot 7 + 0 \cdot 2)^2 + (6 \cdot 0 + 1 \cdot 2 - 0 \cdot 2 - 0 \cdot 7)^2$$
$$= 44^2 + 5^2 + 12^2 + 2^2;$$

we can similarly decompose any integer into a product of primes and then get a representation of it as a sum of four squares, if we know a representation of each prime as a sum of four squares.

Lemma 2. If p is an odd prime, then

$$1 + x^2 + y^2 \equiv 0 \ (\text{mod } p)$$

has a solution with $0 \le x < p/2$ and $0 \le y < p/2$.

Proof. The numbers in

$$S_1 = \left\{ 0^2, \ 1^2, \ 2^2, \ . \ . \ . \ , \left(\frac{p-1}{2}\right)^2 \right\}$$

are distinct (mod p), as are the numbers in

$$S_2 = \left\{ -1 - 0^2, \ -1 - 1^2, \ -1 - 2^2, \ . \ . \ . \ , -1 - \left(\frac{p-1}{2}\right)^2 \right\}.$$

S_1 and S_2 contain together $(p - 1)/2 + 1 + (p - 1)/2 + 1 = p + 1$ numbers. Since there are only p least residues (mod p), we must have one of the numbers in S_1 congruent to one of the numbers in S_2:

$$x^2 \equiv -1 - y^2 \pmod{p}$$

for some x and y, and $0 \le x \le (p - 1)/2$, $0 \le y \le (p - 1)/2$.

For example, take $p = 17$. The numbers in S_1 are 1, 4, 9, 16, 25, 36, 49, 64, and the numbers in S_2 are $-1, -2, -5, -10, -17, -26, -37, -50, -65$. Their least residues (mod 17) are

 0, 1, 4, 9, 16, 8, 2, 15, 13 and 16, 15, 12, 7, 0, 8, 14, 1, 3;

the sets have not one but five elements in common, namely 0, 1, 8, 15, and 16. They give five solutions, three of which are distinct:

$$1 + 0^2 + 4^2 \equiv 1 + 1^2 + 7^2 \equiv 1 + 5^2 + 5^2 \equiv 1 + 7^2 + 1^2 \equiv 1 + 4^2 + 0^2$$
$$\equiv 0 \pmod{17}.$$

To show that every positive integer is the sum of four squares, we use the same method of proof as was used to prove the theorem on integers that are the sum of two squares. We express some multiple of p as a sum of four squares and then construct a smaller multiple of p, also the sum of four squares. Repeating the process often enough will give p as a sum of four squares, and that is all we need. Since $2 = 1^2 + 1^2 + 0^2 + 0^2$, the case $p = 2$ is settled, and we can assume hereafter that p is an odd prime.

Lemma 3. For every odd prime p, there is an integer m, $m < p$, such that

$$mp = x^2 + y^2 + z^2 + w^2$$

has a solution.

Proof. From Lemma 2 we know that there are x and y such that

$$mp = x^2 + v^2 + 1^2 + 0^2$$

for some m. Since $0 \le x < p/2$ and $0 \le y < p/2$, we have

$$mp = x^2 + y^2 + 1 < p^2/4 + p^2/4 + 1 < p^2,$$

so $m < p$.

* **Exercise 1.** From

$$12 \cdot 17 = 204 = 14^2 + 2^2 + 2^2 + 0^2$$

and

$$12 \cdot 17 = 204 = 13^2 + 5^2 + 3^2 + 1^2,$$

find representations of $3 \cdot 17$ as sums of four squares.

Lemma 4. If m and p are odd, $1 < m < p$, and

$$mp = x^2 + y^2 + z^2 + w^2,$$

then there is a positive integer k_1 with $k_1 < m$ such that

$$k_1 p = x_1{}^2 + y_1{}^2 + z_1{}^2 + w_1{}^2$$

for some integers x_1, y_1, z_1, w_1.

Proof. As in the two-squares theorem, we will construct x_1, y_1, z_1, w_1 from x, y, z, w. First note that we can suppose that m is odd. If m is even, then x, y, z, w are all odd, all even, or two are odd and two are even. In any event, we can rearrange the terms so that

$$x \equiv y \pmod 2 \qquad \text{and} \qquad z \equiv w \pmod 2.$$

Hence

$$\frac{mp}{2} = \left(\frac{x-y}{2}\right)^2 + \left(\frac{x+y}{2}\right)^2 + \left(\frac{z-w}{2}\right)^2 + \left(\frac{z+w}{2}\right)^2.$$

If $m/2$ is even, we can repeat the process and express $(m/4)p$ as a sum of four squares. Since $m \ne 0$, eventually we will have an odd multiple of p written as a sum of four squares. For example, from

$$172 \cdot 197 = 81^2 + 85^2 + 93^2 + 107^2$$

we get

$$86 \cdot 197 = 2^2 + 83^2 + 7^2 + 100^2$$

and

$$43 \cdot 197 = 49^2 + 51^2 + 38^2 + 45^2.$$

Now choose A, B, C, and D such that

$$A \equiv x, \qquad B \equiv y, \qquad C \equiv z, \qquad D \equiv w \ (\mathrm{mod}\ m).$$

and such that each lies strictly between $-m/2$ and $m/2$. We can do this since m is odd. It follows that

$$A^2 + B^2 + C^2 + D \equiv x^2 + y^2 + z^2 + w^2 \ (\mathrm{mod}\ m),$$

so

$$A^2 + B^2 + C^2 + D^2 = km$$

for some k. Since

$$A^2 + B^2 + C^2 + D^2 < m^2/4 + m^2/4 + m^2/4 + m^2/4 = m^2,$$

we have $0 < k < m$. (If $k = 0$, then m divides each of x, y, z, and w, so $m^2 \mid mp$. This is impossible, because $1 < m < p$.) Thus

$$m^2 k p = (mp)(km) = (x^2 + y^2 + z^2 + w^2)(A^2 + B^2 + C^2 + D^2),$$

and from Lemma 1 we have

$$m^2 kp = (xA + yB + zC + wD)^2 + (xB - yA + zD - wC)^2$$
$$+ (xC - yD - zA + wB)^2 + (xD + yC - zB - wA)^2.$$

The terms in parentheses are divisible by m:

$$xA + yB + zC + wD \equiv x^2 + y^2 + z^2 + w^2 \equiv 0 \ (\mathrm{mod}\ m),$$
$$xB - yA + zD - wC \equiv xy - yx + zw - wz \equiv 0 \ (\mathrm{mod}\ m),$$
$$xC - yD - zA + wB \equiv xz - yw - zx + wy \equiv 0 \ (\mathrm{mod}\ m),$$
$$xD + yC - zB - wA \equiv xw + yz - zy - wx \equiv 0 \ (\mathrm{mod}\ m).$$

So, if we put

$$x_1 = (xA + yB + zC + wD)/m, \qquad y_1 = (xB - yA + zD - wC)/m,$$
$$z_1 = (xC - yD - zA + wB)/m, \qquad w_1 = (xD + yC - zB - wA)/m,$$

then we have

$$x_1^2 + y_1^2 + z_1^2 + w_1^2 = (m^2 kp)/m^2 = kp.$$

Since $k < m$, the lemma is proved.

Theorem 1. Every positive integer can be written as the sum of four integer squares.

Proof. Suppose that $n = p_1^{e_1} p_2^{e_2} \cdots p_k^{e_k}$. Starting with Lemma 3, repeated application of Lemma 4 gives a solution of $p_i = x^2 + y^2 + z^2 + w^2$ for each i. From Lemma 1, we can write $p_i^{e_i}$ as a sum of four squares for each i. Applying Lemma 1 again (k times), we can get a representation of $p_1^{e_1} p_2^{e_2} \cdots p_k^{e_k}$ as a sum of four squares.

Problems

* 1. Express 31, 37, 41, 43, 47, and 53 as sums of four squares.

2. Express 11, 13, 17, 19, 23, and 29 as sums of four squares.

* 3. From $2 \cdot 17 \cdot 1984 = 67456 = 256^2 + 40^2 + 16^2 + 8^2$, find a representation of $17 \cdot 1984$ as a sum of four squares.

4. From $53 = 7^2 + 2^2 + 0^2 + 0^2$ and Lemma 1, find a representation of $18179 = 7^3 \cdot 53$ as a sum of four squares.

* 5. Use the idea in the proof of Lemma 2 to find solutions of $2 + x^2 + y^2 \equiv 0$ (mod 17).

6. Use the idea in the proof of Lemma 4 applied to $43 \cdot 197 = 49^2 + 51^2 + 38^2 + 45^2$ to get a smaller multiple of 197 written as a sum of four squares.

7. If $8 \mid (x^2 + y^2 + z^2 + w^2)$, show that x, y, z, and w are even.

8. If $n = x^2 + y^2 + z^2 + w^2$, show that by suitable ordering and choices of sign we can get $x + y + z$ to be a multiple of three.

9. If $n = x^2 + y^2 + z^2 + w^2$ and x, y, z, w are nonnegative, show that

$$\min(x, y, z, w) \le n^{1/2}/2 \le \max(x, y, z, w) \le n^{1/2}.$$

10. If t is even and x, y, and z have no common factor, show that $t^2 = x^2 + y^2 + z^2$ is impossible.

20

$x^2 - Ny^2 = 1$

The theory of diophantine equations has not been perfected. There are not many theorems that apply to a really wide class of equations. Usually, special equations are attacked with special methods, and what works for $x^3 + 3xy + y^3 = z^3$ may be worthless for solving $x^3 + 4xy + y^3 = z^3$. (On the other hand, the same method might work for both.) The perfect theorem would be one that would let us look at any diophantine equation and decide whether it had solutions. It would be even better if the theorem would let us decide exactly how many solutions there are, and better yet if it would tell us exactly what they are.

This perfection will never be achieved. There was once hope that it could be approached, though. In 1904, the great German mathematician David Hilbert presented a list of 24 problems which he thought were worthy of the attention of twentieth-century mathematicians. They were in fact worthy, and the search for their solutions has turned up important results; see, for example, Fang [5]. Hilbert's tenth problem was to find, if possible, a method which would allow us to look at a diophantine equation and tell if it had solutions or not. It took more than sixty years to discover that there is no such general method that will work for all diophantine equations, and that it is as useless to look for one as it is to try to trisect angles with straightedge and compass

alone. (Davis [4] has explained, as nontechnically as possible, Matusevic's resolution of the problem.)

To illustrate the state the theory of diophantine equations has reached today, here is one of the most general theorems now known (general in the sense that it applies to a larger class of equations than do other theorems):

Theorem. Let

$$F(x, y) = a_n x^n + a_{n-1} x^{n-1} y + a_{n-2} x^{n-2} y^2 + \cdots + a_0 y^n,$$

and suppose that $F(x, 1) = 0$ has no repeated roots. Then the equation

$$F(x, y) = c,$$

where c is an integer, has only finitely many solutions if $n \geq 3$.

In particular, this theorem says that $ax^n + by^n = c$ has, in general, only finitely many solutions if $n \geq 3$. What if $n < 3$? We have completely analyzed the case $n = 1$ in the section on linear diophantine equations, and, for $n = 2$, we considered a special case in the section on Pythagorean triangles. The general equation when $n = 2$ is too complicated for us to treat here. In this section, we will treat another special case:

$$x^2 - Ny^2 = 1,$$

where N is a positive integer. We will show that if we can find one solution of this equation with $x > 1$, then we can find infinitely many. In fact, if we can find the smallest solution (the one with $x > 1$ as small as possible), then we can find *all* the solutions of the equation.

The equation $x^2 - Ny^2 = 1$ is commonly called Pell's Equation. This is the result of a mistake made by Euler, who called it that. Euler was so eminent that everyone has called it that since. But Pell never solved the equation, and there is even doubt that he could have. The mathematical historian E. T. Bell [3] has written, "Pell mathematically was a nonentity and humanly an egregious fraud. . . . He never even saw the equation." Certainly, frauds and nonentities do not deserve the immortality of having an equation called after them. Fermat not only saw the equation, he was able to solve some portions of it. Hence we will call $x^2 - Ny^2 = 1$ *Fermat's Equation*.

The equation is always satisfied when $x = \pm 1$ and $y = 0$, whatever the value of N. We will call solutions in which either $x = 0$ or $y = 0$ trivial solutions.

* **Exercise 1.** Find, by trial, a nontrivial solution of $x^2 - 2y^2 = 1$, and one of $x^2 - 3y^2 = 1$.

An efficient way to go about finding a nontrivial solution of $x^2 - Ny^2 = 1$ by trial is to make a table of $1 + Ny^2$ for $y = 1, 2, \ldots$ and inspect it for squares.

In solving $x^2 - Ny^2 = 1$, there is no need to consider negative values of N. If $N \le -2$, then it is clear that the equation has only the trivial solutions with $y = 0$, because both terms on the left are nonnegative. For $n = -1$, there are also the solutions $x = 0$, $y = \pm 1$. These are trivial too. Besides supposing N to be positive, we can assume that N is not a square. If it is, then $N = m^2$ for some m, and we have

$$1 = x^2 - m^2 y^2 = (x - my)(x + my).$$

The product of two integers is 1 only when both are 1 or both are -1, and all the solutions can thus be quickly found by solving pairs of linear equations.

* **Exercise 2** (optional). Show that $x = \pm 1$, $y = 0$ are the only solutions.

We will hereafter assume that $N > 0$ and N is not a square. With these assumptions, it is always possible to show that $x^2 - Ny^2 = 1$ always has a solution other than $x = \pm 1$, $y = 0$. We will accept this result on faith and not prove it. There are two methods for proving the existence of a nontrivial solution: one depends on developing the extensive machinery of continued fractions, and the other (first constructed in 1842 by Dirichlet, who improved a proof given by Lagrange in 1766) is not short.

Because

$$x^2 - Ny^2 = (x + y\sqrt{N})(x - y\sqrt{N}),$$

irrational numbers of the form $x + y\sqrt{N}$ are closely connected with solutions of Fermat's Equation. They also have several important properties, which we develop in the following lemmas. We will say that the irrational number

$$\alpha = r + s\sqrt{N}$$

(r and s are integers) *gives a solution* of $x^2 - Ny^2 = 1$ if and only if $r^2 - Ns^2 = 1$. For example, $3 + 2\sqrt{2}$ gives a solution of $x^2 - 2y^2 = 1$ and $8 + 3\sqrt{7}$ gives a solution of $x^2 - 7y^2 = 1$.

Lemma 1. If $N > 0$ is not a square, then

$$x + y \sqrt{N} = r + s \sqrt{N}$$

if and only if $x = r$ and $y = s$.

Proof. If $x = r$ and $y = s$, then clearly $x + y \sqrt{N} = r + s \sqrt{N}$. It is the converse that is important. To prove it, suppose that $x + y \sqrt{N} = r + s \sqrt{N}$ and $y \neq s$. Then

$$\sqrt{N} = \frac{x - r}{s - y}$$

is a rational number. But since N is not a square, \sqrt{N} is irrational. It follows that $y = s$, and this implies $x = r$.

Lemma 2. For any integers a, b, c, d, N,

$$(a^2 - Nb^2)(c^2 - Nd^2) = (ac + Nbd)^2 - N(ad + bc)^2.$$

Proof. Multiply it out. For example, $(2^2 - 3 \cdot 1^2)(7^2 - 3 \cdot 4^2) = (14 + 3 \cdot 1 \cdot 4)^2 - 3(2 \cdot 4 + 1 \cdot 7)^2 = 26^2 - 3 \cdot 15^2$, and this is in fact correct: $(4 - 3)(49 - 48) = 676 - 675$.

Lemma 3. If α gives a solution of $x^2 - Ny^2 = 1$, then so does $1/\alpha$.

Proof. Let $\alpha = r + s \sqrt{N}$. Then we know that $r^2 - Ns^2 = 1$, and we have

$$\frac{1}{\alpha} = \frac{1}{r + s \sqrt{N}} \frac{r - s \sqrt{N}}{r - s \sqrt{N}} = \frac{r - s \sqrt{N}}{r^2 - Ns^2} = r - s \sqrt{N};$$

since $r^2 + N(-s)^2 = 1$, the lemma is proved.

Lemma 4. If α and β give solutions of $x^2 - Ny^2 = 1$, then so does $\alpha\beta$.

Proof. Let $\alpha = a + b \sqrt{N}$ and $\beta = c + d \sqrt{N}$. Then

$$\alpha\beta = (a + b \sqrt{N})(c + d \sqrt{N}) = (ac + Nbd) + (ad + bc) \sqrt{N}$$

and from Lemma 2 we have

$$(ac + Nbd)^2 - N(ad + bc)^2 = (a^2 - Nb^2)(c^2 - Nd^2) = 1,$$

and this shows that $\alpha\beta$ gives a solution.

Exercise 3. Two solutions of $x^2 - 8y^2 = 1$ are $(x, y) = (3, 1)$ and $(17, 6)$. Apply Lemma 4 to find another.

Lemma 5. If α gives a solution of $x^2 - Ny^2 = 1$, then so does α^k for any integer k, positive, negative, or zero.

Exercise 4 (optional). Prove Lemma 5. First show that it is true for all $k \geq 1$ by applying Lemma 4 and induction. Then show that it is true for $k \leq -1$ by applying Lemma 3. Then consider the case $k = 0$.

Lemma 5 shows that if we know one number α, $\alpha > 1$, which gives a solution of $x^2 - Ny^2 = 1$, then we can find infinitely many, namely those given by α^k, $k = 2, 3, \ldots$. The solutions are all different, because $\alpha^{k+1} > \alpha^k$ for all k. For example, $3 + 2\sqrt{2}$ gives a solution of $x^2 - 2y^2 = 1$. So, then, do

$$(3 + 2\sqrt{2})^2 = 17 + 12\sqrt{2}$$

and

$$(3 + 2\sqrt{2})^3 = (17 + 12\sqrt{2})(3 + 2\sqrt{2}) = 99 + 70\sqrt{2}$$

and higher powers of $3 + 2\sqrt{2}$.

Exercise 5. Check that $(3 + 2\sqrt{2})^2$ and $(3 + 2\sqrt{2})^3$ give solutions of $x^2 - 2y^2 = 1$.

Exercise 6. $\alpha = 2 + \sqrt{3}$ gives a solution of $x^2 - 3y^2 = 1$. Find two other nontrivial solutions.

Lemma 6. Suppose that a, b, c, d are nonnegative and that $\alpha = a + b\sqrt{N}$ and $\beta = c + d\sqrt{N}$ give solutions of $x^2 - Ny^2 = 1$. Then $\alpha < \beta$ if and only if $a < c$.

Proof. Suppose that $a < c$. Then $a^2 < c^2$ and because $a^2 = 1 + Nb^2$ and $c^2 = 1 + Nd^2$ we have $Nb^2 < Nd^2$. Because none of b, d, N are negative, it follows that $b < d$. Together with $a < c$, this gives $\alpha < \beta$. To prove the

converse, suppose that $\alpha < \beta$. If $a \geq c$, then $a^2 \geq c^2$. From this follows $b^2 \geq d^2$, which implies $\alpha \geq \beta$. Since this is impossible, we have $a < c$.

Now we are in a position to describe *all* the solutions of $x^2 - Ny^2 = 1$. Consider the set of all real numbers that give a solution of $x^2 - Ny^2 = 1$. Let θ be the smallest number in the set greater than one. Note that Lemma 6 guarantees that there will be such a smallest element, because the members $r + s \sqrt{N}$ of the set can be ordered according to the size of r, which is an integer, and any nonempty set of positive integers contains a smallest element. We will call θ the **generator** for $x^2 - Ny^2 = 1$, and we can now prove

Theorem 1. If θ is the generator for $x^2 - Ny^2 = 1$, then all nontrivial solutions of the equation with x and y positive are given by $\theta^k, k = 1, 2,$
. . . .

Note that the restriction of x and y to positive values loses us nothing essential, because nontrivial solutions come in quadruples

$$\{(x, y), (x, -y), (-x, y), (-x, -y)\},$$

and exactly one solution has two positive elements. Note also that we say nothing about the existence of a generator. It is a fact, as was noted earlier, that such a number can always be found. There is a method for getting θ from the continued fraction expansion of \sqrt{N} by an easy calculation—easy in the sense that a computer would make light work of it. For some values of N, the computation is quite tedious. Of course, a generator can be found by trial, and for a long time, this was the only method available. In the seventh century, the Indian mathematician Brahmagupta said that a person who can within a year solve $x^2 - 92y^2 = 1$ is a true mathematician. Perhaps, and perhaps not; but such a person would at least be a true arithmetician, because the generator is $1151 + 120 \sqrt{92}$. Solution by trial can also be difficult for so innocent-seeming an equation as $x^2 - 29y^2 = 1$; its smallest positive nontrivial solution is $x = 9801$ and $y = 1820$. The equation $x^2 - 61y^2 = 1$ has no positive nontrivial solution until $x = 1766319049, y = 226153980$. You can verify that this is a solution by multiplication, if you wish.

Proof of Theorem 1. Let $x = r, y = s$ be any nontrivial solution of

$x^2 - Ny^2 = 1$ with $r > 0$ and $s > 0$. Let $\alpha = r + s \sqrt{N}$. We want to show that $\alpha = \theta^k$ for some k. We know that $\alpha \geq \theta$ by the definition of generator, so there is a positive integer k such that

$$\theta^k \leq \alpha < \theta^{k+1}.$$

Thus $1 \leq \theta^{-k}\alpha < \theta$. From Lemmas 4 and 5, we know that $\theta^{-k}\alpha$ gives a solution of $x^2 - Ny^2 = 1$. We have defined θ to be the *smallest* number that is greater than one and which gives a nontrivial solution. But $\theta^{-k}\alpha$ is smaller than θ and also gives a solution. Hence $\theta^{-k}\alpha$ gives a trivial solution. Thus $\theta^{-k}\alpha = 1$ or $\alpha = \theta^k$, as we wanted to show.

Problems

* 1. Find the generator for $x^2 - Ny^2 = 1$ when N is 8, 6, 12, or 10.

2. Find the generator for $x^2 - Ny^2 = 1$ when N is 15, 7, 11, or 13.

* 3. Find two positive nontrivial solutions of $x^2 - Ny^2 = 1$ when N is 8, 6, or 63.

4. Find two positive nontrivial solutions of $x^2 - Ny^2 = 1$ when N is 15, 7, or 99.

* 5. Find three nontrivial solutions of $x^2 + 2xy - 2y^2 = 1$.

6. Determine infinitely many solutions of the equation in Problem 5.

*† 7. (a) Show that if $a^2 > b$, then $x^2 + 2axy + by^2 = 1$ has infinitely many solutions if $a^2 - b$ is not a square.
 (b) If $a^2 < b$, show that the equation has only solutions in which $y = 0, 1,$ or -1.
 (c) What happens if $a^2 = b$?

8. (a) Let $a = 2mn$, $b = m^2 - n^2$, and $c = m^2 + n^2$ be the sides of a Pythagorean triangle. Suppose that $b = a + 1$. Show that $(m - n)^2 - 2n^2 = 1$, and determine all such triangles.
 (b) Find the smallest two such triangles.

*† 9. (a) Show that a triangle with sides $2a - 1, 2a, 2a + 1$ has an integer area if and only if $3(a^2 - 1)$ is a square.
 (b) Find three such triangles.

10. (a) Show that a triangle with sides $2a, 2a + 1, 2a + 2$ has a rational area if and only if $3((2a + 1)^2 - 4)$ is a square.
 (b) Show that this is impossible.

11. Show that if $x_1 + y_1 \sqrt{N}$ is the generator for $x^2 - Ny^2 = 1$, then all solutions x_k, y_k can be written in the form

$$2x_k = (x_1 + y_1 \sqrt{N})^k + (x_1 - y_1 \sqrt{N})^k,$$
$$2 \sqrt{N}y_k = (x_1 + y_1 \sqrt{N})^k - (x_1 - y_1 \sqrt{N})^k.$$

12. Show that if $x_1 + y_1 \sqrt{N}$ is the generator of $x^2 - Ny^2 = 1$, then

$$0 < x_1 - y_1 \sqrt{N} < 1.$$

* 13. With the notation used in Problems 11 and 12, what happens to x_k/y_k as k gets larger and larger?

14. If $x_1 + y_1 \sqrt{N}$ is the generator for $x^2 - Ny^2 = 1$ and

$$x_k + y_k \sqrt{N} = (x_1 + y_1 \sqrt{N})^k,$$

$k = 1, 2, \ldots$, show that

$$x_{k+1} = 2x_1 x_k - x_{k-1},$$
$$y_{k+1} = 2x_1 y_k - y_{k-1}.$$

* 15. $10^2 + 11^2 + 12^2 = 13^2 + 14^2$. Find another sum of three consecutive squares equal to a sum of two consecutive squares.

21

Bounds for $\pi(x)$

Prime numbers have always fascinated mathematicians, professional and amateur alike. They appear among the integers, seemingly at random, and yet not quite: there seems to be some order or pattern, just a little below the surface, just a little out of reach. Euler tried to discover "the secret of the primes" but it always eluded him. People have searched for primes, for twin primes, for formulas which would give all the primes or give only primes, and people are still searching today. The largest pair of twin primes—that is, primes whose difference is 2—that is known was discovered in 1972: $76 \cdot 3^{139} - 1$ and $76 \cdot 3^{139} + 1$ are both primes. It is not even known if the number of twin primes is infinite; it is not known whether $n^2 + 1$ is prime infinitely often: the primes do not give up their secrets easily.

In the next section we will look at some formulas for primes, and in this one we will consider the number of primes among the integers 1, 2, . . . , n. Although we may have a hard time telling whether a specific large integer is a prime, we can make quite accurate statements about *how many* integers in a given interval are prime. An actuary for a life insurance company does a similar thing when she makes accurate statements about how many policyholders will die in the coming year, even though she can say nothing about who is doomed. A casino oper-

ater also does a similar thing when he makes accurate statements about how much profit a roulette table will make, even though he can say nothing about how much you will lose. We can often tell what is happening on the whole, even when the details are not clear.

In this section, x will not be restricted to be an integer, nor will y, though other lower case italic letters will continue to denote integers. Let $\pi(x)$ denote the number of primes less than or equal to x.

* **Exercise 1.** What are $\pi(2)$, $\pi(24)$, $\pi(\sqrt{401})$, and $\pi(3\pi)$?

Given a table of primes, counting will give $\pi(x)$ for small values of x:

x	1	10	100	1000	10000	10^5	10^6	10^7
$\pi(x)$	0	4	25	168	1229	9592	78498	664579

It looks as if $\pi(x)$ is increasing, but at a slower rate than x, and it would not be unreasonable to suppose that anyone considering the evidence might sooner or later guess that $\pi(x)$ and $x/\ln x$ are increasing at the same rate. This is indeed so, and the

Prime Number Theorem. As x increases without bound, the ratio of $\pi(x)$ to $x/\ln x$ approaches 1.

was guessed more than 100 years before it was proved. ($\ln x$ denotes the natural logarithm of x.) In fact, for x between 100 and 10^7, the ratio stays between .9 and 1.2, and by the time x is 10^{10} it is 1.048. We will not prove this; instead, we will establish the weaker result that

(1) $$\frac{1}{4} \ln 2 < \frac{\pi(x)}{x/\ln x} < 32 \ln 2$$

for $x \geq 2$. That is, we are showing that the ratio stays between .173 and 22.18. The inequalities in (1) were first proved by Tchebyshev in 1850 with better constants than .173 and 22.18 (they were closer to 1), and he also proved that if the ratio of $\pi(x)$ to $x/\ln x$ approached any limit, it had to be 1. That was the first step on the way toward the proof of the Prime Number Theorem, though Legendre had stated in 1780 that good approximation to $\pi(x)$ was $x/(\ln x - 1.08366)$, and in 1792 Gauss suggested $\int_2^x dy/\ln y$, which fits even better. But Gauss was

not able to suggest a proof, and it was not until 1859 that Riemann attempted a proof which was not adequate, but contained the ideas essential for a complete proof. It was in 1896 that Hadamard and de la Vallée Poussin independently proved the theorem, and work on refinements of it still continues.

The proof of (1) may seem unnatural because some of the preliminary results are proved by induction, and a proof by induction never discloses where the result came from in the first place, and because it may not be clear where the central idea came from, or even what it is. Since it did not occur to Gauss, we do not have to feel bad if we do not see it immediately.

Recall that for $n \geq 1$, $n! = n(n-1) \cdots 3 \cdot 2 \cdot 1$, and that

$$\binom{n}{r} = \frac{n(n-1) \cdots (n-r+1)}{r(r-1) \cdots 1} = \frac{n!}{r!(n-r)!}$$

for $n \geq r \geq 1$. For example, $\binom{5}{3} = \frac{5 \cdot 4}{2 \cdot 1} = \frac{5!}{3! \, 2!}$.

* **Exercise 2.** Evaluate $\binom{6}{3}$ and $\binom{8}{4}$.

Our proof depends on properties of $\binom{2n}{n}$. Its first few values are

n	1	2	3	4	5	6	7
$\binom{2n}{n}$	2	6	20	70	252	1748	17160

They may not seem very interesting, but when they are factored they have some striking properties. Table 1 gives the factorization for $n = 1, 2, \ldots, 20$, and several observations can be made:

1. All the primes between n and $2n$ appear with exponent 1.

2. None of the primes between $2n/3$ and n appear at all.

3. $\binom{2n}{n}$ is always divisible by $n + 1$.

4. Each prime-power is less than $2n$.

The last one is not easy to see, but it is the one we will need. Before we prove that it is true, we need to look at the prime-power decomposition of $n!$ because $\binom{2n}{n} = \frac{(2n)!}{(n!)^2}$, so the power that a prime is

raised to in the prime-power decomposition of $\binom{2n}{n}$ will be the power it is raised to in the prime-power decomposition of $2n!$ minus twice the power it is raised to in the prime-power decomposition of $n!$ The prime-power decomposition of 20! is $2^{18} \cdot 3^8 \cdot 5^4 \cdot 7^2 \cdot 11 \cdot 13 \cdot 17 \cdot 19$, and it is not hard to see why the exponent of 2 is 18. One power of 2 comes from each of the 10 multiples of 2 (2, 4, 6, 8, 10, 12, 14, 16, 18, 20), one more from each of the 5 multiples of 4 (4, 8, 12, 16, 20), one more from each of the 2 multiples of 8 (8, 16), and one from the single multiple of 16 (16).

Table 1. Exponent of p in the prime-power decomposition of $\binom{2n}{n}$

		2	3	5	7	11	13	17	19	23	29	31	37
	1	1											
	2	1	1										
	3	2	0	1									
	4	1	0	1	1								
	5	2	2	0	1								
	6	3	1	0	1	1							
	7	3	1	1	0	1	1						
	8	1	2	1	0	1	1						
	9	1	0	1	0	1	1	1					
	10	2	0	0	0	1	1	1	1				
	11	2	1	0	1	0	1	1	1				
n	12	1	0	0	1	0	1	1	1	1			
	13	2	0	2	1	0	0	1	1	1			
	14	2	3	2	0	0	0	1	1	1			
	15	3	2	1	0	0	0	1	1	1	1		
	16	0	2	1	0	0	0	1	1	1	1	1	
	17	1	3	1	0	1	0	0	1	1	1	1	
	18	1	1	2	1	1	0	0	1	1	1	1	
	19	2	1	2	1	1	0	0	0	1	1	1	1
	20	1	2	1	1	1	1	0	0	1	1	1	1

* **Exercise 3.** What is the exponent of 2 in the prime-power decomposition of 12!? What is the exponent of 3?

The number of multiples of 2 that are less than or equal to 20 is r, where $2 \cdot 1, 2 \cdot 2, \ldots, 2 \cdot r$ are less than 20. That is, r is the largest

integer less than 20/2, or [20/2]. Similarly, the number of multiples of 4 that are less than or equal to 20 is [20/4], and so on:

$$18 = [20/2] + [20/4] + [20/8] + [20/16].$$

In general, we have

Lemma 1. The highest power of p that divides $n!$ is

$$[n/p] + [n/p^2] + [n/p^3] + \cdots .$$

Proof. Each multiple of p less than or equal to n adds one power of p to $n!$, and there are $[n/p]$ such multiples. The multiples of p^2 each contribute an additional power of p, and there are $[n/p^2]$ such multiples. And so on: the additional contribution made by the multiples of p^k is $[n/p^k]$, and hence p to the power $[n/p] + [n/p^2] + [n/p^3] + \cdots$ exactly divides $n!$

As an application of the lemma, we can determine how many zeros there are at the end of 1984! The highest power of 5 that divides 1984! is

$$[1984/5] + [1984/25] + [1984/125] + [1984/625] = 396 + 79 + 15 + 3$$
$$= 493.$$

Since there more than that number of factors of 2 in 1984!, it ends with a string of 493 consecutive zeros.

* **Exercise 4.** How many factors of 2 are there in 1984!?

Lemma 2. The highest power of p that divides $\dbinom{2n}{n}$ is

$$[2n/p] - 2[n/p] + [2n/p^2] - 2[n/p^2] + [2n/p^3] - 2[n/p^3] + \cdots .$$

Proof. Since $\dbinom{2n}{n} = (2n)!/(n!)^2$ we can apply Lemma 1. The numerator contains exactly $[2n/p] + [2n/p^2] + [2n/p^3] + \cdots$ factors of p and the denominator contains exactly $2([n/p] + [n/p^2] + [n/p^3] + \cdots)$ factors of p, so $\dbinom{2n}{n}$ contains exactly their difference.

Note that all these sums of greatest-integer brackets end, since sooner or later $p^k > n$, and so $[n/p^k] = [n/p^{k+1}] = \cdots = 0$.

Lemma 3. For any x, $[2x] - 2[x] \le 1$.

Proof. From the definition of the greatest-integer function, $[2x] \le 2x$ and $[x] > x - 1$, so

$$[2x] - 2[x] < 2x - 2(x - 1) = 2.$$

Since $[2x] - 2[x]$ is an integer, the conclusion follows.

The next lemma is the one which makes the proof work.

Lemma 4. Each prime-power in the prime-power decomposition of $\binom{2n}{n}$ is less than or equal to $2n$.

Proof. Suppose that p^r is in the prime-power decomposition of $\binom{2n}{n}$. Suppose that $p^r > 2n$. Then $[2n/p^r] = [2n/p^{r+1}] = \cdots = 0$, $[n/p^r] = [n/p^{r+1}] = \cdots = 0$, so the sum in Lemma 2 ends after $r - 1$ terms:

$$(2) \qquad r = ([2n/p] - 2[n/p]) + ([2n/p^2] - 2[n/p^2]) + \cdots$$
$$+ ([2n/p^{r-1}] - 2[n/p^{r-1}]).$$

But from Lemma 3, each of the terms in parentheses is at most 1. Thus (2) says that $r \le 1 + 1 + \cdots + 1$ ($r - 1$ terms) or $r \le r - 1$, which is impossible. Thus $p^r \le 2n$.

We next need bounds on $\binom{2n}{n}$.

Lemma 5. For $n \ge 1$, $2^n \le \binom{2n}{n} \le 2^{2n}$.

Proof. We will use mathematical induction. The lemma is true for $n = 1$ because $\binom{2}{1} = 2$. Suppose that it is true for $n = k$. Then

$$\binom{2(k + 1)}{k + 1} = \frac{(2k + 2)!}{((k + 1)!)^2} = \frac{(2k + 2)(2k + 1)(2k!)}{(k + 1)k!(k + 1)k!}$$
$$= \frac{2(2k + 1)}{k + 1} \binom{2k}{k}.$$

On the one hand,

$$\frac{2(2k+1)}{k+1}\binom{2k}{k} < \frac{2(2k+2)}{k+1}\binom{2k}{k} = 4\binom{2k}{k} \le 4\cdot 2^{2k}$$
$$= 2^{2(k+1)},$$

and on the other,

$$\frac{2(2k+1)}{k+1}\binom{2k}{k} > \frac{2(k+1)}{k+1}\binom{2k}{k} = 2\binom{2k}{k} \ge 2\cdot 2^{k}$$
$$= 2^{k+1}.$$

Lemma 6. For $n \ge 2$, $\pi(2n) - \pi(n) \le (2n \ln 2)/\ln n$.

Proof. The idea is that the prime-power decomposition of $\binom{2n}{n}$ contains each prime between n and $2n$. This is because

$$\binom{2n}{n} = \frac{(2n)(2n-1)\cdots(n+1)}{n(n-1)\cdots 1},$$

and none of the primes in the numerator can be canceled by any factor in the denominator. Thus

$$\binom{2n}{n} \ge \prod_{n<p\le 2n} p.$$

But each prime is larger than n, so

$$\prod_{n<p\le 2n} p \ge \prod_{n<p\le 2n} n,$$

which is n multiplied together as many times as there are primes between n and $2n$. That number is $\pi(2n) - \pi(n)$, and so

$$\prod_{n<p\le 2n} n = n^{\pi(2n)-\pi(n)}.$$

So, using Lemma 5, we get

$$2^{2n} \ge \binom{2n}{n} \ge n^{\pi(2n)-\pi(n)}.$$

Taking logarithms of both sides gives $2n \ln 2 \ge (\pi(2n) - \pi(n))\ln n$, which is what we wanted to show.

Lemma 7. For $n \ge 2$, $\pi(2n) \ge n \ln 2/\ln(2n)$.

Proof. From Lemma 4, each prime-power in the prime-power decomposition of $\binom{2n}{n}$ is at most $2n$, and there are at most $\pi(2n)$ prime-powers. Hence $\binom{2n}{n} \le (2n)^{\pi(2n)}$. From Lemma 5, we get $2^n \le (2n)^{\pi(2n)}$, so $n \ln 2 \le \pi(2n) \ln(2n)$, and that is what was to be shown.

Lemma 8. For $r \ge 1$, $\pi(2^{2r}) < 2^{2r+2}/r$.

Exercise 5. Verify that the lemma is true for $r = 1$ and 2.

Proof. We will use mathematical induction. The lemma is true for $r = 1$. Suppose that it is true for $r = k$. Then

$$\pi(2^{2k+2}) = (\pi(2^{2k+2}) - \pi(2^{2k})) + \pi(2^{2k}).$$

From Lemma 6 we have

$$\pi(2^{2k+2}) \le (2 \cdot 2^{2k} \cdot \ln 2)/\ln(2^{2k}) + \pi(2^{2k})$$
$$= 2^{2k}/k + \pi(2^{2k}),$$

and from the induction assumption we get

$$\pi(2^{2k+2}) < 2^{2k}/k + 2^{2k+2}/k = 5 \cdot 2^{2k}/k.$$

We want to show that

$$\pi(2^{2k+2}) < 2^{2k+4}/(k + 1) = 16 \cdot 2^{2k}/(k + 1),$$

and this will be so if $5/k < 16/(k + 1)$. But this is true for $k \ge 1$.

Now we can prove

Theorem 1. For $x \ge 2$,

$$\frac{1}{4} \ln 2 \, (x/\ln x) \le \pi(x) \le 32 \ln 2 \, (x/\ln x).$$

Proof. To get the left-hand inequality, let n be so that $2n \le x < 2n + 2$. We will apply Lemma 7:

$$\pi(x) \ge \pi(2n) \ge \frac{n \ln 2}{\ln(2n)} \ge \frac{n \ln 2}{\ln x} \ge \frac{2n + 2}{4} \frac{\ln 2}{\ln x}$$
$$> \frac{x}{4} \frac{\ln 2}{\ln x}$$

To get the right-hand inequality, choose r so that $2^{2r-2} \le x < 2^{2r}$ and apply Lemma 8:

$$\frac{\pi(x)}{x} \le \frac{\pi(2^{2r})}{2^{2r-2}} < \frac{2^{2r+2}}{2^{2r-2}r} = \frac{16}{r}.$$

But $\ln x < (2r) \ln 2$, so $r > (\ln x)/(2 \ln 2)$, and it follows that

$$\frac{\pi(x)}{x} < \frac{32 \ln 2}{\ln x},$$

which is what we wanted.

Problems

*† 1. (a) What is the highest power of 2 that divides 1984! ?
 (b) For which n does 2^n divide $n!$?

2. Prove by induction a stronger version of part of Lemma 5, namely $\binom{2n}{n} > 2^{2n-1}/\sqrt{n}$ for $n \ge 1$.

† 3. Prove that if $n < p \le 2n$, then p appears in the prime-power decomposition of $\binom{2n}{n}$ to the power 1.

4. Prove that if p is odd and $2n/3 < p \le n$, then p does not appear in the prime-power decomposition of $\binom{2n}{n}$.

† 5. Let p_n denote the nth prime. If $\pi(x) \le ax/\ln x$, show that $p_n \ge \frac{1}{a} n \ln n$.

22

Formulas for Primes

In the earlier days of mathematics, there was a feeling that "function" and "formula" were more or less synonymous. Today, the notion of function is more general, but many of us still feel more comfortable with a function if we have an explicit formula to look at. There is no difference, really, between

$f(n)$ is the largest prime factor of n

and G. H. Hardy's formula

$$f(n) = \lim_{r \to \infty} \lim_{s \to \infty} \lim_{t \to \infty} \sum_{u=0}^{s} [1 - (\cos^2 (u!)^r \pi/n)^{2t}].$$

The first expression is simpler, but perhaps the second lets us feel that we somehow have more control over f. (Some primitive people believe that if you know a man's name, then you have power over him. It is the same principle.)

The importance of formulas is of course not psychological but practical: in general, a formula will let us compute things of interest. Thus the formula above is *less* useful than the verbal description: it obscures what f is, and it does not lend itself to computation. But if we agree that formulas in general are nice things and worth having, then it is reason-

able to search for them. We might ask for a formula for p_n, the nth prime. But the primes are so irregularly scattered through the integers that this is probably beyond all reason. The next best thing would be to have a formula that would produce nothing but primes. The aims of this section are to show that no polynomial formula will work, to exhibit a formula which will (but which is not adapted to computation), to prove that there is a prime between n and $2n$ for all positive integers n, and to use this to get another formula for primes.

The simplest sort of formula to consider is

$$f(n) = an + b.$$

If we found such a function that gave nothing but primes, then we would have an arithmetic progression, with difference a, consisting entirely of primes. Looking through tables of primes, we can find various arithmetic progressions of primes, but none of infinite length: for example,

3, 5, 7;

7, 37, 67, 97, 127, 157;

199, 409, 619, 829, 1039, 1249, 1459, 1669, 1879, 2089.

But no infinite arithmetic progression can be made up entirely of primes, as we now show. Suppose that $p \nmid a$, so $(a, p) = 1$ and hence there is an integer r such that $ar \equiv -b \pmod{m}$. Then

$$a(r + kp) + b \equiv ar + b \equiv 0 \pmod{p}$$

for $k = 0, 1, \ldots$, so every pth term of the sequence is divisible by p. The longest arithmetic progression known that consists entirely of primes is $223092870n + 2236133941$ for $n = 0, 1, \ldots, 15$ [7], and it is not known if there exist arbitrarily long arithmetic progressions of primes.

After seeing that a sequence $\{an + b\}$ cannot consist entirely of primes, it is natural to ask whether the sequence can contain infinitely many primes. The answer to this is given by Dirichlet's Theorem: If $(a, b) = 1$, then the sequence $\{an + b\}$ contains infinitely many primes. For example, among the members of the sequence $\{4n + 1\}$ are the primes 5, 13, 17, 29, 37, 41, \ldots , and among $\{12n + 7\}$ are 7, 19, 31, 43, 67, \ldots ; Dirichlet's Theorem says that we will never come to a last prime in either sequence. The condition $(a, b) = 1$ is clearly necessary: $\{6n + 3\}$ contains only one prime and $\{6n + 4\}$ contains none. Dirichlet's great achievement was in showing that the condition was

also sufficient. The proof of this theorem is not at all easy, and we will not attempt it.

However, some special cases are easy. Consider $\{3n + 2\}$, suppose that there are only finitely many primes in it, and call them p_1, p_2, . . . , p_k. Let $N = p_1 p_2 \cdots p_k$. If $N \equiv 1 \pmod 3$, then $N + 1 \equiv 2 \pmod 3$ and thus must have at least one prime divisor congruent to 2 (mod 3) (otherwise N would be congruent to 1 (mod 3)). But $N + 1 \equiv 1 \pmod{p_k}$, so whatever prime divisor it has is not one of p_1, p_2, . . . , p_k. If $N \equiv 2 \pmod 3$, then $N + 3 \equiv 2 \pmod 3$, and $N + 3$ must have a prime divisor congruent to 2 (mod 3). But $N + 3 \equiv 3 \pmod{p_k}$, so its prime divisor is not one of p_1, p_2, . . . , p_k either. This is impossible so the assumption that there were only finitely many primes in the sequence was wrong.

No polynomial can have only prime values, either. If $f(n) = a_0 n^k + a_1 n^{k-1} + \cdots + a_k$, and if r is such that $f(r) \equiv 0 \pmod p$ for some p, then $f(r + mp) \equiv f(r) \equiv 0 \pmod p$ for $m = 1, 2, \ldots$. Just as with arithmetic progressions, if one term is divisible by p, then every pth term from there on is also divisible by p. The champion quadratic polynomial for having consecutive prime values is $n^2 + n + 41$, found by Euler. It has a prime value for $n = -40, -39, \ldots, 39$. However, $40^2 + 40 + 41 = 41^2$, and every forty-first term in the sequence $\{n^2 + n + 41\}$ is divisible by 41. The analogue of Dirichlet's Theorem for higher-degree polynomials would be that $\{a_0 n^k + a_1 n^{k-1} + \cdots + a_k\}$ contains infinitely many primes if a_0, a_1, \ldots, a_k have no common factor. No such theorem has been proved, and it is not even known if $n^2 + n + 41$ is prime infinitely often, though it seems unlikely that this should not be so.

On the other hand, we can construct a polynomial that assumes as many consecutive prime values as we want, because it can be shown that it is always possible to make a polynomial of degree d take on $d + 1$ arbitrarily assigned values. For example, if

$$60f(x) = 7x^5 - 85x^4 + 355x^3 - 575x^2 + 418x + 180,$$

then we have

n	0	1	2	3	4	5
$f(n)$	3	5	7	11	13	17

A similar polynomial could be constructed to take on 81 consecutive prime values, but it would be of degree 80.

After giving up on polynomials, it would be natural to try exponential functions. For example, if

$$f(n) = [(3/2)^n],$$

then $f(n)$ is prime for $n = 2, 3, 4, 5, 6, 7$ (the values of the function are 2, 3, 5, 7, 11, and 17), but $f(8) = 25$, and the next prime in the sequence does not come until $f(21) = 4987$. No one has proved that a formula like $f(n) = [\theta^n]$ cannot always give a prime. Nor is it known whether $[\theta^n]$ can be prime infinitely often. Such questions seem hopelessly difficult.

Nevertheless, there do exist functions, expressible as a simple formula, that always represent primes. We will prove, partly, a striking result of Mills [11]:

Theorem 1. There is a real number θ such that $[\theta^{3^n}]$ is a prime for all n, $n = 1, 2, \ldots$.

As we shall see, this theorem contains less than meets the eye, and it should not seem nearly so striking after we finish the proof. The proof gives a construction for θ, but the construction depends on being able to recognize arbitrarily large primes. If we could recognize arbitrarily large primes, we would have no need of the formula.

In the proof we will use two theorems from analysis.

Theorem. If a sequence $u_1, u_2, u_3, \ldots, u_n, \ldots$ is bounded above and nondecreasing, then it has a limit, θ, as n increases without bound.

That is, if there is a number M such that $u_n < M$ for all n and $u_n \le u_{n+1}$ for all n, $n = 1, 2, \ldots$, then there is a number θ such that the difference between θ and u_n becomes arbitrarily small as n increases without bound. We will not prove this theorem, or the next.

Theorem. If a sequence $v_1, v_2, v_3, \ldots, v_n, \ldots$ is bounded below and nonincreasing, then it has a limit, ϕ, as n increases without bound.

We will write

$$\lim_{n \to \infty} u_n = \theta \qquad \text{and} \qquad \lim_{n \to \infty} v_n = \phi$$

and read "the limit of u_n as n approaches infinity equals θ," and a corresponding statement for v_n.

Proof of Theorem 1. The proof depends on the following theorem: there is an integer A such that if $n > A$, then there is a prime p such that

$$(1) \qquad\qquad n^3 < p < (n + 1)^3 - 1.$$

We will not prove this but we will use it to determine a sequence of primes that will in turn determine θ. Let p_1 be any prime greater than A, and for $n = 1, 2, \ldots$, let p_{n+1} be a prime such that

$$(2) \qquad\qquad p_n^3 < p_{n+1} < (p_n + 1)^3 - 1.$$

Such a prime exists for each n on account of (1). Let

$$(3) \qquad u_n = p_n^{3^{-n}} \qquad \text{and} \qquad v_n = (p_n + 1)^{3^{-n}},$$

$n = 1, 2, \ldots$. We see that as n increases, u_n increases, because from (2),

$$(4) \qquad u_{n+1} = p_{n+1}^{3^{-n-1}} > (p_n^3)^{3^{-n-1}} = p_n^{3^{-n}} = u_n.$$

Furthermore, $\{v_n\}$ is a decreasing sequence, because from (2),

$$(5) \quad v_{n+1} = (p_{n+1} + 1)^{3^{-n-1}} < ((p_n + 1)^3 - 1 + 1)^{3^{-n-1}} = (p_n + 1)^{3^{-n}} = v_n.$$

It is clear from (3) that $u_n < v_n$. Hence, because of (5),

$$u_n < v_n < v_{n-1} < \cdots < v_1,$$

so $u_n < v_1$ for all n. In the same way, from (4) we have

$$v_n > u_n > u_{n-1} > \cdots > u_1,$$

so $v_n > u_1$ for all n. Thus $\{u_n\}$ is an increasing sequence of numbers that is bounded above by v_1. It follows that $\{u_n\}$ has a limit. Call it θ. Also, $\{v_n\}$ is a decreasing sequence of numbers that is bounded below by u_1. Hence $\{v_n\}$ has a limit too. Call it ϕ. Since $u_n < v_n$ for all n, it follows that $\theta \le \phi$. In fact, since $\{u_n\}$ increases and $\{v_n\}$ decreases, we have

$$u_n < \theta \le \phi < v_n$$

for all n; thus

$$u_n^{3^n} < \theta^{3^n} \le \phi^{3^n} < v_n^{3^n}$$

for all n. But from the definitions of u_n and v_n,

$$u_n^{3^n} = p_n \qquad \text{and} \qquad v_n^{3^n} = p_n + 1.$$

Thus

$$p_n < \theta^{3^n} < p_n + 1.$$

This locates θ^{3^n} between two consecutive integers, and so

$$[\theta^{3^n}] = p_n,$$

a prime, for all n.

From the construction, we see that knowledge of θ and knowledge of all the primes is essentially equivalent, so that the theorem gives us nothing that we did not have before, except perhaps pleasure at seeing a clever idea neatly worked out. The theorem would be important only if we could discover what θ is by some method independent of all the primes, and this is not likely.

To prepare for getting another formula for primes, we will derive a result of independent interest, *Bertrand's Theorem:*

Theorem 2. For all $n \geq 2$ there is a prime p such that $n < p < 2n$.

Proof. Just as in the last section, we will need properties of $\binom{2n}{n}$, and we will need to recall the binomial formula

$$(6) \quad (x + y)^n = x^n + \binom{n}{1} x^{n-1}y$$

$$+ \binom{n}{2} x^{n-2}y^2 + \cdots + \binom{n}{n-1} xy^{n-1} + y^n,$$

true for any x and y. We will assume that for some n there are no primes p such that $n < p < 2n$, or what is the same thing for $n < p \leq 2n$, since $2n$ is not a prime when $n \geq 2$. We will show that this implies that $n < 2788$, so that the theorem is true for $n \geq 2788$. By checking the cases $n = 1, 2, \ldots, 2787$, we can then complete the proof.

First we will show that for $n \geq 2$,

$$(7) \qquad \prod_{p \leq n} p \leq 2^{2n}.$$

This is not a strong inequality—for $n = 10$ it asserts only that $210 \leq 1048576$—but it is all that we need. The binomial expansion of $(1 + 1)^{2m+1}$ is

$$(1 + 1)^{2m+1} = 1 + \binom{2m + 1}{1} + \cdots + \binom{2m + 1}{m} + \binom{2m + 1}{m + 1}$$

$$+ \cdots + \binom{2m + 1}{2m} + 1$$

$$\geq \binom{2m+1}{m} + \binom{2m+1}{m+1}.$$

The two terms are both equal to

(8)
$$\frac{(2m+1)(2m)\cdots(m+2)}{m(m-1)\cdots 1},$$

so $2^{2m+1} \geq 2 \binom{2m+1}{m}$ or

(9)
$$2^{2m} \geq \binom{2m+1}{m}.$$

Also $\binom{2m+1}{m}$ is divisible by each prime p such that $m + 1 < p \leq 2m + 1$, as inspection of (8) shows. Thus

(10)
$$\prod_{m+1<p\leq 2m+1} p \leq \binom{2m+1}{m} \leq 2^{2m}.$$

Now we can prove (7) by mathematical induction. It is true for $n = 2$. Suppose that it is true for all $n \leq k$. If k is odd, then $k + 1$ is even, and

(11)
$$\prod_{p\leq k+1} p = \prod_{p\leq k} p \leq 2^{2k} < 2^{2(k+1)}.$$

If k is even, say $k = 2m$, then

(12)
$$\prod_{p\leq k+1} p = \left(\prod_{p\leq m+1} p \right) \left(\prod_{m+1<p\leq 2m+1} p \right)$$
$$\leq 2^{2(m+1)} 2^{2m} = 2^{4m+2} = 2^{2(k+1)},$$

where the induction assumption was used on the first product and (10) on the second. (11) and (12) complete the induction.

Remember that we have assumed that $\binom{2n}{n} = N$ has no prime divisors p such that $n < p \leq 2n$. But

$$N = \frac{(2n)(2n-1)\cdots(n+1)}{n(n-1)\cdots 1},$$

so if $2n/3 < p \leq n$, then p is a factor in the denominator, and since $2p > 4n/3 \geq n + 1$, $2p$ is a factor in the numerator. The two p's cancel, and since $3p > 2n$, there are no more factors of p in the numerator. Thus all of the prime divisors of N are at most $2n/3$, so

(13)
$$\prod_{p|N} p \leq \prod_{p\leq 2n/3} p \leq 2^{4n/3};$$

the last inequality comes from applying (7).

We will use (13) to get an upper bound for N. From Lemma 4 of Section 21, we know that each prime power in the prime-power decomposition of N is at most $2n$. So, if p appears in the prime-power decomposition to be a power greater than 1, then $p^2 \leq 2n$ and $p \leq \sqrt{2n}$. There are at most $\sqrt{2n}$ such primes, and since each prime power is at most $2n$, their contribution to the prime-power decomposition is at most $(2n)^{\sqrt{2n}}$. All of the other primes appear to the power 1, and from (13), their product is at most $2^{4n/3}$. Thus

$$(14) \qquad \binom{2n}{n} \leq 2^{4n/3}(2n)^{\sqrt{2n}}.$$

On the other hand, mathematical induction shows that

$$(15) \qquad \binom{2n}{n} \geq \frac{2^{2n}}{2n}$$

$$\left(\text{because } \binom{2n+2}{n+2} = \frac{(2n+2)(2n+1)}{(n+2)(n+1)} \binom{2n}{n} \geq \frac{2(2n+1)}{n+1} \frac{2^{2n}}{2n}\right.$$

$$\left. = \frac{2n+1}{2n} \frac{2^{2n+2}}{2n+2} \geq \frac{2^{2n+2}}{2n+2}\right),$$

so combining (14) and (15) we get

$$\frac{2^{2n}}{2n} \leq 2^{4n/3}(2n)^{\sqrt{2n}},$$

and this does not hold if n is too large. Taking logarithms, we get

$$2n \ln 2 - \ln 2n \leq (4n/3) \ln 2 + \sqrt{2n} \ln 2n$$

or

$$(2n/3) \ln 2 \leq (\sqrt{2n} + 1) \ln 2n \leq (\sqrt{2n} + \sqrt{2n}) \ln 2n$$
$$= 2\sqrt{2} \sqrt{n} \ln 2n$$

or

$$\sqrt{n} \leq \frac{3\sqrt{2} \ln 2n}{\ln 2}.$$

Since \sqrt{n} increases more rapidly than $\ln 2n$, this inequality is false for n sufficiently large: in fact, for $n > 2787$. The sequence of primes

$$2, 3, 5, 7, 13, 23, 43, 83, 163, 317, 631, 1259, 2503, 9973,$$

each less than twice the one before, shows that for $n \leq 2787$ there is always at least one prime between n and $2n$.

We can use this to get another formula for primes like the one in Theorem 1, but which does not depend on an unproved theorem [16].

Choose any p_1 and for $n = 1, 2, \ldots,$ let p_{n+1} be a prime such that

(16)
$$2^{p_n} < p_{n+1} < 2^{p_n+1};$$

such a prime exists because of Theorem 2. Let

(17)
$$u_n = \log^{(n)} p_n, \qquad v_n = \log^{(n)}(p_n + 1),$$

where $\log^{(1)} k = \log_2 k$ and $\log^{(n)} k = \log_2(\log^{(n-1)} k)$. Taking logarithms to the base 2 in (16) gives

$$p_n < \log_2 p_{n+1} < p_n + 1,$$

and since $p_{n+1} + 1 \le 2^{p_{n+1}}$, we have

$$p_n < \log^{(1)} p_{n+1} < \log^{(1)}(p_{n+1} + 1) \le p_n + 1.$$

If we take logarithms to the base 2 of the preceding inequalities n times, we have

$$u_n < u_{n+1} < v_{n+1} \le v_n.$$

So, as in the proof of Theorem 1, $\theta = \lim_{n \to \infty} u_n$ and $\phi = \lim_{n \to \infty} v_n$ exist. Let $\exp^{(1)} k = 2^k$ and $\exp^{(n)} k = 2^{\exp^{(n-1)} k}$. Then from $u_n < \theta < v_n$ we get $\exp^{(n)} u_n < \exp^{(n)} \theta < \exp^{(n)} v_n$, or

$$p_n < \exp^{(n)} \theta < p_n + 1,$$

and we have $[\exp^{(n)} \theta] = p_n$ for all n. Restating this, we have proved

Theorem 3. There exists a real number θ such that $[2^\theta]$, $[2^{2^\theta}]$, $[2^{2^{2^\theta}}]$, \ldots are all prime.

Various other formulas for primes have been devised. For example, we can use Wilson's Theorem to get a formula for $\pi(x)$. Recalling that

$$\frac{(n-1)! + 1}{n}$$

is an integer if n is prime and is not an integer if n is composite,

$$\cos^2 \pi \left(\frac{(n-1)! + 1}{n} \right) = f(n)$$

is 1 if n is prime, and less than 1 if n is composite. Thus

$$\pi(x) = \sum_{2 \le n \le x} [f(n)].$$

Another striking result, also based on Wilson's Theorem, appeared in 1976 [6]: the set of prime numbers is identical with the set of

positive values taken on by the following polynomial of degree 25 in the 26 variables x_1, x_2, \ldots, x_{26}:

$$(x_{11} + 2)\{1 - [x_{23}x_{26} + x_8 + x_{10} - x_{17}]^2$$
$$- [(x_7x_{11} + 2x_7 + x_{11} + 1)(x_8 + x_{10}) + x_8 - x_{26}]^2$$
$$- [2x_{14} + x_{16} + x_{17} + x_{26} - x_5]^2$$
$$- [16(x_{11} + 1)^3(x_{11} + 2)(x_{14} + 1)^2 + 1 - x_6^2]^2$$
$$- [x_5^3(x_5 + 2)(x_1 + 1)^2 + 1 - x_{15}^2]^2$$
$$- [(x_1^2 - 1)x_{25}^2 + 1 - x_{24}^2]^2 - [16x_{18}^2x_{25}^4(x_1^2 - 1) + 1 - x_{21}^2]^2$$
$$- [(x_1 + x_{21}^2(x_{21}^2 - x_1)^2 - 1)(x_{14} + 4x_4x_{25})^2 + 1 - (x_{24} + x_3x_{21})^2]^2$$
$$- [x_{14} + x_{12} + x_{22} - x_{25}]^2 - [(x_1^2 - 1)x_{12}^2 + 1 - x_{13}^2]^2$$
$$- [x_1x_9 + x_{11} + 1 - x_{12} - x_9]^2$$
$$- [x_{16} + x_{12}(x_1 - x_{14} - 1) + x_2(2x_1x_{14} + 2x_1 - x_{14}^2 - 2x_{14} - 2) - x_{13}]^2$$
$$- [x_{17} + x_{25}(x_1 - x_{16} - 1) + x_{19}(2x_1x_{16} + 2x_1 - x_{16}^2 - 2x_{16} - 2) - x_{24}]^2$$
$$- [x_{26} + x_{16}x_{12}(x_1 - x_{16}) + x_{20}(2x_1x_{16} - x_{16}^2 - 1) - x_{16}x_{13}]^2\}.$$

To have a set of problems following that would be anticlimactic, and so the text will stop here.

23

Additional Problems

There is no quicker way of developing mathematical power and maturity than by doing problems—*not* routine exercises, in which you substitute numbers in some formula, or mimic some worked-out example in a book, or differentiate yet another function, but *problems* where you must apply what you know but do not know in advance how it will be applied. There is no growth in repeating exercises, but there is growth in looking at a problem of a sort never before seen and asking what ideas could possibly be applied to it, then trying to apply them. It is in considering problems that mathematicians, and mathematics, develop. Of course, there is no growth without some pain, error, and frustration, and there will be many problems—perhaps most—which will defeat you. But that is the way of it: no matter how good you are, there will be problems you cannot solve; and no matter how bad you are, there will be problems that are simple for you.

Some problems follow. They are listed by the section of this book in which they could have appeared. In addition, there are 100 miscellaneous problems, arranged very roughly in order of difficulty, without regard to subject.

Section 1

* 1. Find two solutions of $299x + 247y = 52$.

2. (a) Prove that $(a, b) = (a, c) = 1$ implies $(a, bc) = 1$.
 (b) Prove that $(a, b) = 1$ and $c \mid a$ imply $(c, b) = 1$.

3. (a) Prove that if $a \mid c$, $b \mid c$, and $(a, b) = d$, then $ab \mid cd$.
 (b) Prove that if $(a, c) = 1$ and $(b, c) = d$, then $(ab, c) = d$.

4. Prove that if d is odd, $d \mid (a + b)$, and $d \mid (a - b)$, then $d \mid (a, b)$.

5. Prove that if $p \mid (ra - b)$ and $p \mid (rc - d)$ for some r, then $p \mid (ad - bc)$.

6. If a, b, and c are positive integers, prove that $a \mid c$, $c \mid b$, and $(a, b) = 1$ together imply that $a = 1$.

* 7. Student A says, "I've been looking for a half hour for n such that n and $n + 20$ have a greatest common divisor of 7 and I haven't found one. I think I'll program it for the computer." Student B says, "The computer won't find one, either." How did B know that?

8. Let (a, b, c) denote the greatest common divisor of a, b, and c.
 (a) Prove that $(a, b, c) = ((a, b), c)$.
 (b) If $(a, b) = (b, c) = (a, c) = 1$, prove that $(a, b, c) = 1$.
 (c) Show that the converse of (b) is false.

* 9. Is it true that $(k, n + rk) = d$ for all integers r if and only if $(k, n) = d$?

10. Show that if $10 \mid (3^m + 1)$ for some m, then $10 \mid (3^{m+4n} + 1)$ for all positive integers n.

Section 2

* 1. Complete and prove the

 Theorem. An integer n is a kth power if and only if every exponent _____ .

2. (a) Find the smallest positive integer n such that $n + 1$, $n + 2$, $n + 3$, and $n + 4$ are all composite.
 (b) How many consecutive composite integers follow $k! + 1$?

* 3. Define the *least common multiple* of a and b (written $[a, b]$) to be the smallest integer m such that $a \mid m$ and $b \mid m$.
 (a) Find $[12, 30]$ and $[pq, 2p^2]$, where p and q are distinct odd primes.
 (b) Show that $a = p_1^{e_1} p_2^{e_2} \cdots p_k^{e_k}$ and $b = p_1^{f_1} p_2^{f_2} \cdots p_k^{f_k}$ imply

 $$[a, b] = p_1^{g_1} p_2^{g_2} \cdots p_k^{g_k},$$

 where $g_i = \max(e_i, f_i)$ $(e_i \geq 0$ and $f_i \geq 0)$, $i = 1, 2, \cdots, k$.
 (c) Prove that $ab = a, b$.

4. Let $2, 3, \ldots, p_n$ be the first n primes. Let $N = 2 \cdot 3 \cdot \ldots \cdot p_n$. If $N = ab$, prove that $a + b$ has a prime divisor greater than p_n.

† 5. Establish the following test for primes. If n is odd, greater than 5, and there exist relatively prime integers a and b such that

$$a - b = n \qquad \text{and} \qquad a + b = p_1 p_2 \cdots p_k$$

(where p_1, p_2, \ldots, p_k are the odd primes $\leq n^{1/2}$), then n is prime.

Section 3

* 1. How many different ways can thirty nickels, dimes, and quarters be worth $5?

2. The following problem is at least 400 years old: Find the number of men, women, and children in a company of 20 if together they pay $20, each man paying $3, each woman $2, and each child 50¢.

3. A says, "We three have $100 altogether." B says, "Yes, and if you had six times as much and I had one-third as much, we three would still have $100." C says, "It's not fair. I have less than $30." Who has what?

4. Anna took 30 eggs to market and Barbara took 40. Each sold some of her eggs at 5 cents per egg and later sold the remainder at the same lower price (in cents per egg). Each received the same amount of money. What is the smallest amount that they could have received?

5. A man sold n cows for $$n$ per cow. With the proceeds, he bought an odd number of sheep for $10 each and a pig for less than $10. How much did the pig cost?

Section 4

1. Prove that if $d \mid m$ and $a \equiv b \pmod{m}$, then $a \equiv b \pmod{d}$.

2. True or false? $a \equiv b \pmod{m}$ implies $a^2 \equiv b^2 \pmod{m^2}$.

*† 3. (a) What can a square be, modulo 9?
 (b) Is 314,159,267,144 a square?

4. (a) What is the largest integer with ten distinct digits that is divisible by 9?
 (b) What is the largest integer with eight distinct digits that is divisible by 9?

5. Suppose that $ar \equiv b \pmod{m}$ and $br \equiv a \pmod{m}$ for some r. Show that $a^2 \equiv b^2 \pmod{m}$ and find a nontrivial example of such $a, b, r,$ and m.

6. Show that $a^5 \equiv a \pmod{10}$ for all a.

* 7. Find an integer n such that $n \equiv 1 \pmod{2}$, $n \equiv 0 \pmod{3}$, and $n \equiv 0 \pmod{5}$. Can you find infinitely many?

8. Show that no triangular number has as its last digit 2, 4, 7, or 9. (A triangular number is one of the form $n(n + 1)/2$.)

9. Prove that if p is a prime and p divides no one of $a_1, a_2, \ldots, a_{p-1}$ nor any of their differences, then $a_1, a_2, \ldots, a_{p-1}$ are congruent (mod p) to 1, 2, $\ldots, p - 1$ in some order.

10. If $n = 31,415,926,535,897$, then let

$$f(n) = 897 - 535 + 926 - 415 + 031 = 904.$$

Induce a definition for f, and prove that if $7 | f(n)$, then $7 | n$; if $11 | f(n)$, then $11 | n$; and if $13 | f(n)$, then $13 | n$. Check 118,050,660 for divisibility by 2, 3, 5, 7, 11, and 13.

Section 5

* 1. Solve $9x \equiv 4$ (mod 2401).

2. Find the smallest positive integer m such that $2^5 | m$ and $3^5 | m + 1$.

*† 3. For which positive integers does

$$kx \equiv 1 \ (\text{mod } k(k + 1)/2)$$

have a solution?

4. If $(s, m) = 1$, denote the solution of $sx \equiv r$ (mod m) by r/s. Prove that

$$(r/s)(t/u) \equiv (rt)/(su) \ (\text{mod } m)$$

and that

$$(r/s) + (t/u) \equiv (ru + st)/(su) \ (\text{mod } m).$$

5. Consider the system

$$x \equiv a_i \ (\text{mod } m_i), \qquad i = 1, 2, \ldots, k,$$

where the moduli are relatively prime in pairs. Let

$$M_i = (m_1 m_2 \cdots m_k)/m_i, \qquad i = 1, 2, \ldots, k.$$

Let s_i denote the solution of $M_i x \equiv 1$ (mod m_i), $i = 1, 2, \ldots, k$. Show that

$$s = a_1 s_1 M_1 + a_2 s_2 M_2 + \cdots + a_k s_k M_k$$

satisfies each of the congruences in the system.

Section 6

† 1. Let a' denote the solution of $ax \equiv 1$ (mod p), $a = 1, 2, \ldots, p - 1$.
 (a) Show that it is always true that $(ab)' \equiv a'b'$ (mod p).
 (b) Show that it is not always true that $(a + b)' \equiv a' + b'$ (mod p).

2. It is known [12] that if p is prime, then

$$a^p(p - 1)! \equiv a(p - 1)(\text{mod } p)$$

for all a. Show that this implies both that

$$a^{p-1} \equiv 1 \ (\text{mod } p) \ \text{if } (a, p) = 1$$

and

$$(p - 1)! \equiv -1 \ (\text{mod } p).$$

† 3. (a) Prove that if $r! \equiv (-1)^r \ (\text{mod } p)$, then

$$(p - r - 1)! \equiv -1 \ (\text{mod } p).$$

(b) Find an example of such a p and r.

4. (a) Show that

$$(k + 1)^p - k^p \equiv 1 \ (\text{mod } p),$$

for $k = 0, 1, \ldots$.
(b) Derive Fermat's Theorem from this.

5. A composite n such that $n | (2^n - 2)$ is called a **pseudoprime**. There are infinitely many, and the smallest two are 341 and 561. Verify that 561 is a pseudoprime.

6. A composite n such that $n | (a^n - a)$ for all a is called an **absolute pseudoprime**. The smallest absolute pseudoprime is 561. Show that 341 is not an absolute pseudoprime by verifying that $341 \nmid (11^{341} - 11)$.

† 7. Calculate $(2p - 1)! \ (\text{mod } p^2)$ for some values of p and guess a theorem.

8. Prove that if $2^m \not\equiv 1 \ (\text{mod } p)$, then

$$1^m + 2^m + \cdots + (p - 1)^m \equiv 0 \ (\text{mod } p).$$

† 9. If p is a prime, show that $p + 2$ is prime if and only if

$$4((p - 1)! + 1) + p \equiv 0 \ (\text{mod } p + 2).$$

10. If p is an odd prime, and if $(a, p) = 1$, $n | p - 1$, and $a \equiv c^n \ (\text{mod } p)$, prove that

$$p | (a^{(p-1)/n} - 1).$$

Section 7

1. Cardano was the first to mention $d(n)$ when, in 1537, he said that if p_1, p_2, \ldots, p_k are distinct primes, then

$$d(p_1 p_2 \cdots p_k) - 1 = 1 + 2 + 2^2 + \cdots + 2^{k-1}.$$

Verify that this is so.

2. Descartes noted, in 1638, that

$$\sigma(p^n) - p^n = \frac{p^n - 1}{p - 1}$$

for $n = 1, 2, \ldots$. Verify that this is so.

* 3. Let us call n a *practical number* if every positive integer less than or equal to n is a sum of distinct divisors of n.
 (a) Show that 12 is practical.
 (b) Show that 10 is not.
 (c) Discover a practical number greater than 12.
 (d) Show that every power of two is practical.

4. Let $f(n)$ denote the number of positive odd divisors of n.
 (a) Make a table of f for $n = 2, 3, 4, \ldots, 15$.
 (b) Show that $f(2^n p^m) = m + 1$ (p an odd prime).
 (c) Guess a formula for $f(2^n p_1^{e_1} p_2^{e_2} \cdots p_k^{e_k})$ (p_i an odd prime).
 (d) Prove it by induction on k.

* 5. Suppose that $n + \sigma(n)$ is divisible by 3.
 (a) If n is prime, show that $n \equiv 1 \pmod 6$.
 (b) Can n be the square of a prime?
 (c) If $n = pq$, p and q odd primes, show that one of them is 3 and the other is congruent to 5 (mod 6).

Section 8

1. Numerologists have noted that prime years often coincide with or presage ill-fortune. The reason, they say, is that prime numbers are the most deficient of numbers. They point out that since 1910, prime years have been 1913 (World War I next year), 1931 and 1933 (the Great Depression), 1949 and 1951 (Korean War), and 1973 (oil embargo and recession). Since 1993, 1997, and 1999 will all be prime years, some numerologists are looking forward to events as large as the end of the world in the last decade of this century. Test the numerologists' hypothesis by seeing which of the last few years were the most abundant, both in the numerological and non-numerological sense.

2. If p^e is one of an amicable pair, show that $\sigma(p^e) = \sigma \left(\dfrac{p^e - 1}{p - 1} \right)$.

3. Show that pq ($pq \neq 6$) is deficient.

4. Show that p^e, $e \geq 1$, is deficient.

† 5. Prove that except for the pairs 3, 5 and 5, 7, the number between every pair of twin primes is abundant.

6. For what values of a is $3 \cdot 5^a \cdot 7$ abundant?

* 7. For what values of a is $2^a \cdot 11$ abundant?

8. For what values of a is $2^a p$, p odd, abundant?

9. Show that if $(m, n) = 1$ and n is abundant, then mn is abundant.

10. Let $n = 2^k(2^{k+1} - 1)$. If $2^{k+1} - 1$ is composite, show that n is abundant.

11. Suppose that $n = 2^p(2^{p+1} - 1)$ is perfect and $q < 2^{p+1} - 1$ is prime. Show that $2^p q$ is abundant.

12. Show that if m and n are an amicable pair,

$$\left(\sum_{d|m} d^{-1} \right)^{-1} + \left(\sum_{d|n} d^{-1} \right)^{-1} = 1.$$

† 13. Show that $\sigma(1 + p) < 1 + p + p^2$, and use this to conclude that p^2 can never be one of an amicable pair.

14. If $d > 0$, $d|n$ and $(d, n/d) = 1$, then d is called a *unitary divisor* of n.
 (a) What are the unitary divisors of 120? Of 360?
 (b) Which integers are such that their only divisors are unitary divisors?
 (c) If $n = p_1^{e_1} p_2^{e_2} \cdots p_k^{e_k}$, how many unitary divisors has n?
 (d) If

$$\sum d = 2n,$$

where the sum is taken over the unitary divisors, d, of n, then n is called a *unitary perfect* number. Find two such numbers.

15. Here is Euler's original proof of Theorem 2. Fill in any missing details.
 Let $n = 2^k m$ be perfect, m odd. The sum, $(2^{k+1} - 1)\sigma(m)$, of the divisors of n must equal $2n$. Thus

$$m/\sigma(m) = (2^{k+1} - 1)/2^{k+1},$$

a fraction in lowest terms. Hence $m = (2^{k+1} - 1)c$ for some integer c. If $c = 1$, then $m = 2^{k+1} - 1$ must be prime, because $\sigma(m) = 2^{k+1}$. If $c > 1$, then $\sigma(m) \geq m + (2^{k+1} - 1) + c + 1$. Thus

$$\frac{\sigma(m)}{m} \geq \frac{2^{k+1}(c + 1)}{m} > \frac{2^{k+1}}{2^{k+1} - 1},$$

a contradiction.

Section 9

1. Show that $\phi(n + k) = 2\phi(n)$ has at least one solution for each k:

 if $(k, 6) = 1$, take $n = 2k$;
 if $(k, 6) = 2$, take $n = k$;
 if $(k, 6) = 3$, take $n = k/3$;
 if $(k, 6) = 6$, take $n = k$.

2. Show that the geometric mean of n and $\phi(n)$ is not an integer if $n = p^a$, $a \geq 2$.

* 3. Find all n such that $\phi(2n) = \phi(3n)$.

4. If n is composite, prove that $\phi(n) \le n - n^{1/2}$.

† 5. Show that if $6 \mid n$, then $\phi(n) \le n/3$.

6. Let $n = dm$. Show that there are $\phi(m)$ positive integers less than n whose greatest common divisor with n is d.

* 7. Let $\phi^{(2)}(n) = \phi(\phi(n))$, $\phi^{(3)}(n) = \phi(\phi^{(2)}(n))$, and so on. Let $e(n)$ denote the smallest integer such that

$$\phi^{(e(n))}(n) = 2.$$

Calculate $e(n)$ for
(a) $n = 3, 4, 5, 6, 7, 8, 9$.
(b) $n = 2^k, k \ge 2$.
(c) $n = 3^k, k \ge 1$.
(d) $n = 2^k 3^j, k \ge 1, j \ge 1$.

8. If $(m, n) = p$, how is $\phi(mn)$ related to $\phi(m)\phi(n)$?

† 9. Show that $\phi(n) = n/3$ if and only if $n = 2^k 3^j$ for some positive integers k and j.

10. Prove the theorem that you guessed in Problem 9 of Section 9.

11. Suppose we know that $\phi(1) = 1$,

$$\phi(np) = p\phi(n) \quad \text{if} \quad p \mid n, \text{ and}$$
$$\phi(np) = (p - 1)\phi(n) \quad \text{if} \quad p \nmid n.$$

Deduce from these formula for $\phi(n)$.

12. If a and b are positive integers, p and q are primes, and $p > q$, show that $\phi(p^a) = \phi(q^b)$ implies $a = 1$.

*† 13. Calculate

$$\sum_{d \mid n} (-1)^{n/d} \phi(d)$$

for
(a) $n = 12, 13, 14, 15, 16$.
(b) $n = p, p$ an odd prime.
(c) $n = 2^k, k \ge 1$.
(d) $n = p^k, k \ge 1$ and p an odd prime.
(e) Guess a theorem.

14. Find all n such that $4 \nmid \phi(n)$.

* 15. Find a positive integer $k, k > 7$, such that $\phi(n) = 2k$ is impossible.

Section 10

* 1. If $(a, m) \ne 1$, for what values of t is $a^t \equiv 1 \pmod{m}$?

2. Which of the integers from 1975 to 1985 have primitive roots?

* 3. Find the smallest prime which has 10 for a primitive root.

4. Show that 2 is not a primitive root of 31.

* 5. Student A says, "Look. These five pages of computation show that $1983^{1983} \equiv 1 \pmod{1024}$. Isn't that amazing?" Student B says, after a glance, "No, it's not amazing—it's wrong." How did B know that without checking the computations?

6. Show that if $(n, p - 1) = 1$, then $x^n \equiv a \pmod{p}$ has exactly one solution.

† 7. If n is a positive integer, define a^{-n} by

$$a^{-n} \equiv r \pmod{p} \qquad \text{if and only if} \qquad ra^n \equiv 1 \pmod{p}.$$

Prove that if m and n are positive integers, $a^m a^{-n} \equiv a^{m-n} \pmod{p}$.

8. With the notation of Problem 7, prove that $(a^m)^{-n} \equiv a^{-mn} \pmod{p}$.

* 9. It is well known that $a^t = k$ if and only if $t = \log_a k$. Let us define the *index* of an integer (mod p) analogously. If g is a primitive root of p, then

$$g^t \equiv k \pmod{p} \qquad \text{if and only if} \qquad t \equiv \text{ind}_g k \pmod{p - 1}.$$

Calculate $\text{ind}_2 k \pmod{19}$ for $k = 1, 2, \ldots, 18$.

10. With the notation of Problem 9, prove that

$$\text{ind}_g ab \equiv \text{ind}_g a + \text{ind}_g b \pmod{p - 1}.$$

† 11. With the notation of Problem 9, prove that

$$\text{ind}_g a^n \equiv n \, \text{ind}_g a \pmod{p - 1}.$$

12. Use the results of Problems 9 and 10 to solve $13x \equiv 16 \pmod{19}$.

13. Use the results of Problems 9 and 11 to solve $x^{13} \equiv 16 \pmod{19}$.

14. Suppose that $\text{ind}_2(p - 1) = r$. Show that $\text{ind}_2(p - 2) = r + 1$. What is x if $\text{ind}_2 x = r + 2$?

† 15. Prove the following generalization of Wilson's Theorem:

$$\prod_{\substack{n=1 \\ (n,m)=1}}^{m} n \equiv \begin{cases} -1 \pmod{m} & \text{if } m \text{ has a primitive root} \\ 1 \pmod{m} & \text{otherwise.} \end{cases}$$

Section 11

* 1. Solve $x^2 + x + 1 \equiv 0 \pmod{5}$, $x^2 + x \equiv 0 \pmod{5}$, and $x^2 + x - 1 \equiv 0 \pmod{5}$.

2. Find quadratic congruences (mod 11) with solutions 5, 6; 5, 7; and 9, 10.

* 3. Find the quadratic residues (mod 31).

4. Use Euler's Criterion to calculate (mod 23) 2^{11}, 3^{11}, 4^{11}, 5^{11}, 22^{11}, and 21^{11}.

* 5. For which of $p = 3, 5, 7, 11, 13,$ and 17 is $x^2 \equiv -2$ (mod p) solvable?

6. (a) How many solutions does $x^2 \equiv 1$ (mod 16) have?
 (b) But don't quadratic congruences have two solutions or no solutions? What's wrong?

* 7. Does $x^2 \equiv 53$ (mod 97) have a solution? Does $x^2 \equiv 97$ (mod 53)?

8. Suppose that $p \equiv 1$ (mod 4). Then $x^2 \equiv -1$ (mod p) has two solutions: call them i and $-i$. Prove or disprove: $a + bi \equiv 0$ (mod p) implies $a \equiv b \equiv 0$ (mod p).

*† 9. If $p \equiv 7$ (mod 8) and $(p - 1)/2$ is prime, is $(p - 1)/2$ a quadratic residue (mod p)?

10. Suppose that $p = q + an^2$, where p and q are odd primes. Is it true that $(a/p) = (a/q)$?

Section 12

*† 1. Theorem 3 had for its hypothesis: "If p and $4p + 1$ are both primes . . ." Examples are 7, 29 and 13, 53, and in each of these the third number in the sequence $(4 \cdot 29 + 1 = 117$ and $4 \cdot 53 + 1 = 213)$ is not prime. Would a computer search help in discovering longer sequences such that $p, q = 4p + 1, r = 4q + 1, s = 4r + 1, . . .$ are all primes?

2. (a) Show that if $p \mid (n^2 + 2an + b)$ for some n, then $((a^2 - b)/p) = 1$.
 (b) Which primes can divide $n^2 + 2n + 2$?

*† 3. (a) Show that $n^2 + (n + 1)^2 + (n + 2)^2 = m^2$ is impossible.
 (b) Show that $n^2 + (n + 1)^2 + \cdots + (n + k)^2 = m^2$ is impossible whenever $1^2 + 2^2 + \cdots + k^2$ is a quadratic nonresidue (mod $k + 1$).
 (c) What are the first three such values of k?

4. (a) Suppose that $p \geq 5$ is prime. Show that -3 is a quadratic residue (mod p) if $p \equiv 1$ or 7 (mod 12) and a nonresidue if $p \equiv 5$ or 11 (mod 12).
 (b) Suppose that p is an odd prime, $p \neq 3$, and $p \nmid a$. Suppose that $x^3 \equiv a$ (mod p) has a solution r. Then

 $$(x - r)(x^2 + xr + r^2) \equiv 0 \text{ (mod } p).$$

 Show that $x^2 + xr + r^2 \equiv 0$ (mod p) has two solutions different from r if and only if $p \equiv 1$ or 7 (mod 12).
 (c) If $p \geq 5$, show that the number of distinct nonzero cubic residues (mod p) is

 $$\begin{array}{llll} p - 1 & \text{if} & p \equiv 5 & \text{or} \quad 11 \text{ (mod 12)} \\ (p - 1)/3 & \text{if} & p \equiv 1 & \text{or} \quad 7 \text{ (mod 12)}. \end{array}$$

† 5. If $p = 2^m + 1$ is a prime, show that every quadratic nonresidue of p is a primitive root of p.

Section 13

*† 1. (a) Find a base b such that $aa_b = 34$.
 (b) Find a base b such that $aaa_b = 1842$.
 (c) Show that it is impossible for $aaa_b = bb_c$ for positive a, b, and c.

2. What is the smallest positive integer n such that the last digit of n is 1 in the bases b, $b + 1$, and $b + 2$?

3. (a) Show that $121_3 = 4^2$, $121_4 = 5^2$, and $121_5 = 6^2$.
 (b) Guess and prove a theorem.
 (c) Evaluate 169_b in base 10 ($b \geq 10$).

4. Show that 111_b is not a perfect square in any base b, $b = 2, 3, \ldots$.

† 5. Prove that every positive odd integer can be represented in the form

$$n = d_0 + d_1 \cdot 2 + d_2 \cdot 2^2 + \cdots + d_k \cdot 2^k,$$

where $d_i = 1$ or $-1, i = 0, 1, \ldots, k$, but the representation is not unique.

Section 14

1. Show that the last digit of x^n, $n = 2, 3, \ldots$
 (a) is 0 if $x = 6$
 (b) is 4 if $x = \chi$
 (c) is x if $x = 3, 5, 7, 8$, or ϵ and n is odd.

2. With which digits can a prime end?

3. If n is an even perfect number, $n \neq 6$, show that the last digit in its duodecimal representation is four.

4. (a) Let n be an integer written in the base do, and let m be its reversal. Show that $\epsilon \mid (n - m)$.
 (b) Generalize to any base b.

† 5. Using the fact that $1001 = 7 \cdot 11 \cdot 17$, develop tests for the divisibility of an integer by 7, 11, and 17.

Section 15

* 1. Student A says, "With enormous labor, I have divided 1 by 31415, and the decimal has a period of 15707, half the maximum. It's a good thing I didn't have to go all the way out to 31414 places." Student B says, "You made a mistake somewhere. Again." How did B know that without checking the calculations?

2. In the hexadecimal system (base 16, with digits $1, 2, \ldots, 9, A, B, \ldots, F$) find the expansions of $1/2, 1/3, \ldots, 1/F$.

* 3. In which bases will the decimal representation of 7/60 terminate?

4. Let $(.a_2 a_3 a_4 \ldots)_{\text{square}}$ be defined by

$$a_2/2^2 + a_3/3^2 + a_4/4^2 + \cdots , 0 \le a_i < i^2.$$

Does every number between 0 and 1 have such a representation? Is it unique?

* 5. Both 1/13 and 1/14 have period 6. Find the next pair of consecutive integers such that their reciprocals have the same period.

Section 16

† 1. Prove that if the sum of two consecutive integers is a square, then the smaller is a leg and the larger is a hypotenuse of a Pythagorean triangle.

2. (a) Given a, how would you find b such that $a^2 + b^2$ is a square?
 (b) Carry out such a procedure for $a = 13$ and $a = 14$.

† 3. Let the generators of a Pythagorean triangle be consecutive triangular numbers. Show that one side of the triangle generated is a cube.

4. Note that $4^2 - 3^2 = 7$, $12^2 - 5^2 = 7 \cdot 17$, and $8^2 - 15^2 = -7 \cdot 23$. Show that $a^2 + b^2 = c^2$ and $(7, abc) = 1$ imply $7 \mid (a^2 - b^2)$.

† 5. Show how to determine all primitive Pythagorean triangles whose area is numerically equal to k times its perimeter, k a positive integer.

Section 17

1. Given that $x^4 + y^4 = z^4$ has no solutions in integers, prove that it has no solutions in rational numbers.

2. Show that $x^{4n} + y^{4n} = z^{4n}$, $n = 1, 2, \ldots$, has no nontrivial solutions.

3. Suppose that we can show that $x^p + y^p = z^p$ has no nontrivial solutions for any odd prime p. Conclude from this and Problem 2 that $x^n + y^n = z^n$ has no solutions for any $n \ge 3$.

4. Show that $x^p + y^p = z^p$ implies $p \mid (x + y - z)$.

† 5. Show that $x^{p-1} + y^{p-1} = z^{p-1}$ has no nontrivial solutions unless $p \mid xyz$.

6. Show that

$$x^3 + y^3 + z^3 + x^2 y + y^2 z + z^2 x + xyz = 0$$

has no nontrivial solutions.

*† 7. Show that there are infinitely many nontrivial solutions of

$$x^n + y^n = z^{n+1}$$

for any $n \ge 1$, namely those given by

$$x = (ac)^{r^n}, \qquad y = (bc)^{r^n}, \qquad z = c^s,$$

where

$$c = a^{rn^2} + b^{rn^2},$$

a and b are arbitrary, and r and s are chosen to satisfy

$$rn^2 + 1 = (n + 1)s.$$

Does the last equation have infinitely many solutions in positive integers r, s?

8. Find a solution of $x^4 + y^4 = z^5$.

* 9. Find solutions to $x^n + y^n = z^{n-1}$.

10. Show that $x^n + y^n = z^m$ has nontrivial solutions if $(n, m) = 1$.

Section 18

1. Show that $4 \mid (x^2 + y^2 + z^2)$ implies that x, y, and z are even.

2. Verify that if $n \equiv 7 \pmod 8$, then n cannot be written as a sum of three squares.

† 3. From Problems 1 and 2, show that $n = 4^e(8k + 7)$ for some nonnegative e and k implies that $n = x^2 + y^2 + z^2$ is impossible.

4. A mathematician said in 1621 that $3n + 1$ is not the sum of three squares if $n = 8k + 2$ or $32k + 9$ for some k. Show that he was wrong.

† 5. Show that not every positive integer n can be written $n = x^2 - y^2$ for some integers x, y.

6. Show that every positive integer n can be written

$$n = x^2 + y^2 - z^2$$

for some integers x, y, z.

† 7. Show that if n is the sum of two triangular numbers, then $4n + 1$ is a sum of two squares.

8. Which integers can be written as the sum of two squares of *rational* numbers?

*† 9. Which of $2 + 1$, $2 \cdot 3 + 1$, $2 \cdot 3 \cdot 5 + 1$, $2 \cdot 3 \cdot 5 \cdot 7 + 1$, . . . can be written as a sum of two squares?

10. Is it possible to mimic the proof of Theorem 2 to prove a generalization: if $(-w/p) = 1$, then there are integers x and y such that $p = x^2 + wy^2$?

Section 19

*† 1. Express 5,724,631 as a sum of four squares.

2. Tabulate the number of different representations of 1, 2, 3, . . . as sums of four squares, and see if any theorem suggests itself.

3. Which integers can be written as a sum of exactly four nonzero squares?

4. Fill in any missing details in Euler's original proof of Lemma 2. Suppose that $(-1/p) = 1$. Then there is an integer x such that $1 + x^2 \equiv 0 \pmod p$. So, suppose that $(-1/p) = -1$ and that the lemma is false. Then $1 + 1 - 2 = 0$ shows that $(-2/p) = -1$, and hence that $(2/p) = 1$. Then $1 + 2 - 3 = 0$ shows that $(-3/p) = -1$ and that $(3/p) = 1$. In this way, $1, 2, \ldots, p - 1$ are all quadratic residues (mod p).

† 5. If $n > 0$ and $8 \mid n$, show that n is not the sum of fewer than eight squares of odd integers.

Section 20

*† 1. If $x_k + y_k \sqrt{2} = (3 + 2\sqrt{2})^k$, calculate $(x_k/y_k) - \sqrt{2}$ for $k = 1, 2, 3, 4$. ($\sqrt{2} = 1.414213562 \ldots$)

2. (a) Show that $x^2 - Ny^2 = -1$ has no solutions if $N \equiv 3 \pmod 4$.
 (b) Show that if x_1, y_1 is a solution of $x^2 - Ny^2 = -1$ with $x_1 > 1$, then u_k, v_k, $k = 1, 2, \ldots$ are solutions of $x^2 - Ny^2 = 1$, where

$$u_k + v_k \sqrt{N} = (x_1 + y_1 \sqrt{N})^{2k}.$$

† 3. Show that if $x^2 - Ny^2 = k$ has one solution, then it has infinitely many.

4. Let $x_{n+1} = x_n + ry_n$ and $y_{n+1} = x_n + y_n, n = 1, 2, \ldots$. Show that $x_k^2 - ry_k^2$ takes on only two different values if $x_0^2 - ry_0^2 = 2$.

* 5. Apply Problem 14 of Section 20 to extend the sequence of rational approximations to $\sqrt{2}$ found in Problem 1 to one more term.

Miscellaneous Problems

1. Prove that $6 \mid (n^3 - n)$ for all positive integers n.

2. Prove that the sum of three consecutive cubes is always divisible by nine.

3. Show that if $a + b$ is even, then $24 \mid ab(a^2 - b^2)$.

4. Show that $(2^n + (-1)^{n+1})/3$ is an odd integer for $n \geq 1$.

* 5. A man came into a post office and said to the clerk, "Give me some 13-cent stamps, one-fourth as many 9-cent stamps, and enough 3-cent stamps so this $5 will pay for them all." How many stamps of each kind did he buy?

6. Construct a stamp problem for the future: a man bought some 25-cent stamps, one-fourth as many 20-cent stamps, and enough 10-cent stamps to make the total exactly $n. What values of n, $n = 1, 2, \ldots$ give a unique solution in positive integers?

* 7. Stamp problems can be endlessly varied: in a state with no sales tax, bottles of scotch sell for $7 each, bottles of rum for $6 each, and bottles of

vodka for $5 each. The same disagreeable man as in Problem 5 came into a liquor store and said, "Give me some bottles of scotch, half as many of rum, and some of vodka. Here's $40. Keep the change—there won't be any, har, har, har." What did he get?

8. Show that the sequence 5, 12, 19, 26, . . . contains no term of the form 2^a or $2^a - 1$.

† 9. Induce a theorem from the following facts:

$$3^2 + 4^2 = 5^2,$$
$$10^2 + 11^2 + 12^2 = 13^2 + 14^2,$$
$$21^2 + 22^2 + 23^2 + 24^2 = 25^2 + 26^2 + 27^2,$$
$$36^2 + 37^2 + 38^2 + 39^2 + 40^2 = 41^2 + 42^2 + 43^2 + 44^2.$$

10. A palindrome is a number that reads the same backward as forward, such as 3141413.
 (a) How many two-digit palindromes are there?
 (b) How many three-digit ones?
 (c) How many k-digit ones?

11. Let us say that an integer is *powerful* if and only if $p^2 \mid n$ whenever $p \mid n$. Prove that $n = r^2 s^3$ for integers r and s where s is square-free—that is, no square divides s.

12. If a and b are positive integers, let us say that a divides b **weakly** (or, that a is a **weak divisor** of b), written $a \int b$, if and only if $p \mid a$ implies $p \mid b$ for primes p.
 (a) Find examples of integers a and b such that $a > b$ and $a \int b$.
 (b) Prove that $a \mid b$ implies $a \int b$.
 (c) Prove that $a \int b$ and $b \int c$ implies $a \int c$.
 (d) Prove that $ab \int c$ implies $a \int c$ and $b \int c$.
 (e) Prove that $ac \int bc$ and $(a, c) \int (b, c)$ imply $a \int b$.
 (f) Prove that $a \int b$ implies $(a, c) \int (b, c)$ for all positive integers c.
 (g) Prove that if there are integers m and n such that $a^n \int b^m$, then $a \int b$.
 (h) Prove that $a \int c$ and $b \int c$ imply $ab \int c$.
 (i) Which of (c) to (h) are false for ordinary divisibility of positive integers? Give examples.

† 13. Construct a formula for f such that $f(n)$ is 1/2 if n is even and 1 if n is odd.

14. Induce a theorem from the following data:

$$1^4 + 1^4 + 2^4 = 2 \cdot 3^2, \qquad 1^4 + 2^4 + 3^4 = 2 \cdot 7^2,$$
$$1^4 + 3^4 + 4^4 = 2 \cdot 13^2, \qquad 2^4 + 3^4 + 5^4 = 2 \cdot 19^2,$$
$$2^4 + 5^4 + 7^4 = 2 \cdot 39^2, \qquad 3^4 + 4^4 + 7^4 = 2 \cdot 37^2.$$

15. Use mathematical induction to prove that $6^n \equiv 1 + 5n \pmod{25}$.

16. A paper was written recently to show that $x^3 + 117y^3 = 5$ has no solutions [10]. It used the theory of algebraic numbers. Show that the equation has no solutions by considering it (mod 9).

* 17. Find the smallest integer n such that n is positive and $25 \mid n$, $36 \mid n + 1$, and $49 \mid n + 2$.

18. Show that
$$\sum_{d \mid n} 1/d = \sigma(n)/n.$$

*† 19. When Ann is half as old as Mary will be when Mary is three times as old as Mary is now, Mary will be five times as old as Ann is now. Neither Ann nor Mary may vote. How old is Ann?

20. Show that if
$$a = r^2 - 2rs - s^2,$$
$$b = r^2 + s^2,$$
$$c = r^2 + 2rs - s^2$$

for some integers r, s, then a^2, b^2, c^2 are three squares in arithmetic progression.

21. Pascal once wrote that he had discovered that the difference of the cubes of any two consecutive integers, less one, is six times the sum of all the positive integers less than or equal to the smaller one. Prove that he was right.

22. Show that if $n = a^2 + b^2 = c^2 + d^2$ with $b \neq d$, then
$$n = \frac{((a - c)^2 + (b - d)^2)((a + c)^2 + (b - d)^2)}{4(b - d)^2},$$

and hence that if n can be written as a sum of two squares in two distinct ways, then n is composite.

* 23. Using the result of Problem 20, factor
(a) $533 = 23^2 + 2^2 = 22^2 + 7^2$,
(b) $1073 = 32^2 + 7^2 = 28^2 + 17^2$.

24. Show that for any a and d such that $(3d, m) = 1$, the system
$$ax + (a + d)y \equiv a + 2d \pmod{m}$$
$$(a + 3d)x + (a + 4d)y \equiv a + 5d \pmod{m}$$

has the same solution.

25. If $n = a^2 + b^2 + c^2$, with a, b, c nonnegative, show that
$$(n/3)^{1/2} \leq \max(a, b, c) \leq n^{1/2}.$$

26. Gauss proved that a regular polygon with m sides can be constructed with ruler and compass if $m = 2^a n$, where a is an integer and $n = 1$ or n is a product of distinct primes of the form $2^k + 1$. List all the regular polygons with fewer than 40 sides that can be constructed with ruler and compass.

*† 27. $1^2 + 2^2 = 3^2 - 2^2$. $2^2 + 3^2 = 7^2 - 6^2$. $3^2 + 4^2 = 13^2 - 12^2$.
$4^2 + 5^2 = 21^2 - 20^2$. What happens in general?

28. It is known that $8k + 3 = x^2 + y^2 + z^2$ has a solution for any $k \geq 0$.
 (a) Show that x, y, and z are odd.
 (b) Deduce that k is equal to a sum of three triangular numbers.

† 29. $(1/3)^2 + (2/3) = (1/3) + (2/3)^2$. Is this astonishing?

30. $(5 + 5/24)^{1/2} = 5 + (5/24)^{1/2}$; are there other numbers like that?

*† 31. (a) Find all a and p such that $a^2 \equiv -2$, $a^3 \equiv 3$, and $a^4 \equiv 4 \pmod{p}$ with $0 < a < p$.
 (b) Show that $a^2 \equiv 2$, $a^3 \equiv 3$, and $a^4 \equiv 4 \pmod{p}$ is impossible with $0 < a < p$.

32. If $p \geq 5$ is prime, show that $p^2 + 2$ is composite.

† 33. (a) Suppose that $0 \leq m < 121$. If $210n + m$ is prime, show that m is prime.
 (b) Generalize: prove that if $P_k = p_1 p_2 \cdots p_k$ (p_i denotes the ith prime) and $0 \leq m < p_{k+1}^2$, then $P_k n + m$ prime implies m prime.

34. (a) Find all primes p such that $3p + 1$ is a square.
 (b) Find all primes p such that $3p + 2$ is a square.

† 35. Fermat was sometimes as blind as the rest of us. He wrote to Roberval, "Permit me to ask you for the demonstration of this proposition which I frankly confess I have not yet been able to find, although I am assured that it is true. If a, b are integers, and if

(1) $$a^2 + b^2 = 2(a + b)x + x^2,$$

both x and x^2 are irrational." Help Fermat out: show that the equation is not satisfied for any integer x, unless $a = b = 0$.

36. Apply the rational root theorem to complete the demonstration in Problem 35. That is, show that if (1) has no integer roots, then it has no rational roots. What conditions on a and b guarantee that x^2 is irrational?

37. If $n = (6m + 1)(12m + 1)(18m + 1)$, show that $n - 1$ is divisible by $36m$.

38. Is .123456789101112131415 . . . a rational number?

† 39. From the factor table, those numbers n such that $10n + 1$, $10n + 3$, $10n + 7$, and $10n + 9$ are all prime are $n = 1, 10, 19, 82, 148, 187, 208,$ What do those numbers have in common and why?

40. $36 | 10116$ and $36 | 100116$. Coincidence? Does $36 | 1000116$? Almost any pocket calculator can show that is true, but no pocket calculator can show that $36 | 10^n + 116$ for all $n \geq 2$. You do it.

*† 41. Bhaskara, in the *Lilivati* (1150 AD) asked for four numbers whose sum is equal to the sum of their squares, and gave the pretty answer $1/3 + 2/3 + 3/3 + 4/3 = (1/3)^2 + (2/3)^2 + (3/3)^2 + (4/3)^2$. Find all solutions in positive integers of $1/d + 2/d + c/d = (1/d)^2 + (2/d)^2 + (c/d)^2$.

42. It is fairly striking that $6^4 \equiv 4 \pmod{19}$ and $6^5 \equiv 5 \pmod{19}$. Show that a necessary condition for $a^k \equiv k \pmod{p}$ and $a^{k+1} \equiv k + 1 \pmod{p}$ is that $p | (k + 1)^k - k^{k+1}$, and find another example.

† 43. Show that any proper divisor of an even perfect number is a deficient number.

44. Show that $[2x] = [x] + [x + 1/2]$ and $[3x] = [x] + [x + 1/3] + [x + 2/3]$. Can you prove a general theorem?

45. Show that the last digit of every Fermat number $2^{2^n} + 1$, is seven, $n = 2, 3, \ldots$.

46. Find all of the solutions of $x^2 + y^2 = (2x + y)^2$, and show that $x^2 + y^2 = ((x + y)/2)^2$ is impossible in positive integers.

† 47. In [9] there is a report on a search for numbers with the property that the sum of their squares is unchanged if their digits are reversed. The example given, $17^2 + 26^2 + 87^2 + 49^2 = 94^2 + 78^2 + 62^2 + 71^2$, is wrong. Can you reconstruct the true equality for which this is a misprint?

48. Show that if

$$\begin{vmatrix} 1 & a & b \\ a & 1 & c \\ b & c & 1 \end{vmatrix} = 1,$$

then $a = b = c = 0$.

*† 49. February 1968 had five Thursdays. What other years before 2100 will have such Februaries?

50. Several years ago today, a man borrowed an integer number of dollars at a normal rate of simple interest. Today he repaid the loan in full with $204.13. How much did he borrow, how long ago, and at what rate of interest?

*† 51. (a) Show that $111 \ldots 11$ (n digits, all ones) is composite if n is composite.
 (b) Is the converse true?

52. Find all 17 solutions of

$$\frac{1}{x} + \frac{1}{y} = \frac{1}{6}$$

in integers (positive or negative).

*† 53. How many solutions does

$$\frac{1}{x} + \frac{1}{y} = \frac{1}{N}$$

have for a given positive integer N?

54. Prove that the sum of twin primes (that is, $2p + 2$, where p and $p + 2$ are both primes) is divisible by 12 if $p > 3$.

*† 55. (a) If p is an odd prime, how many elements in the sequence

$$1 \cdot 2, \ 2 \cdot 3, \ 3 \cdot 4, \ \ldots, p(p + 1)$$

are relatively prime to p?
 (b) If p is an odd prime, how many elements in the sequence

$$1 \cdot 2, \ 2 \cdot 3, \ 3 \cdot 4, \ \ldots, p^2(p^2 + 1)$$

are relatively prime to *p*?

(c) Any guesses for a general theorem?

56. Show that $n^2 + (n + 1)^2 = km^2$ is possible only when -1 is a quadratic residue (mod k).

57. Let $f(x)$ be nonnegative and nondecreasing for $x \geq 0$. Let us say that f is semimultiplicative if and only if $f(nm) \geq f(n)f(m)$ for all positive integers m and n.
 (a) Show that $f(n) = n^k$, k a positive integer, is semimultiplicative.
 (b) Prove that the product of two semimultiplicative functions is semi-multiplicative.
 (c) Show that if $g(x)$ is nonnegative for $x \geq 0$, then $n^{g(n)}$ is semimultiplicative.

58. If a and b are positive integers, then a^2 ends with an even number of zeros, and $10b^2$ ends with an odd number of zeros. Hence $10b^2 = a^2$ is impossible in nonzero integers, and it follows that $10^{1/2}$ is irrational.
 (a) Adapt the above proof to integers in the base b (b not a square) to show that $b^{1/2}$ is irrational.
 (b) Show that $b^{1/m}$ (b not an mth power, $m = 3, 4, \ldots$) is irrational by the same argument.

† 59. Consider the following lists:

List 1				List 2				List 4			
1	9	17	25	2	10	18	26	4	12	20	28
3	11	19	27	3	11	19	27	5	13	21	29
5	13	21	29	6	14	22	30	6	14	22	30
7	15	23	31	7	15	23	31	7	15	23	31

List 8				List 16			
8	12	24	28	16	20	24	28
9	13	25	29	17	21	25	29
10	14	26	30	18	22	26	30
11	15	27	31	19	23	27	31

Pick a number from 1 to 31—any number—and see which of the above lists it appears in. If you add the numbers of the lists in which the number appears, you will get the number you picked. Why does this trick work?

60. Let $P_n = p_1 p_2 \cdots p_n$ and $a_k = 1 + kP_n$, $k = 0, 1, \ldots, n - 1$, where the p's are the primes 2, 3, 5, 7, . . . in ascending order. Show that $(a_i, a_j) = 1$ if $i \neq j$.

† 61. If n is an even perfect number, show that the harmonic mean of the divisors of n is an integer.

62. (a) Show that the least residues of 1, 7, 7^2, . . . , 7^5 (mod 36) are in arithmetic progression.
 (b) What are the least residues of 1, 7, 7^2, 7^3, . . . (mod 216)?

† 63. Show that $2r^2 - 3$ is never a square, $r = 2, 3, \ldots$.

64. In 1494, L. Pacioli said that 134217727 was prime. Show that he was wrong.

* 65. Suppose that a, b, and c have no common factor. Show that solutions to

$$ax + by + cz = 1$$

are given by

$$x = rt + crm + nb/d,$$
$$y = st + csm - na/d,$$
$$z = u - dm,$$

where m and n are arbitrary integers, r and s are such that $ar + bs = d = (a, b)$, and t and u are such that $dt + cu = 1$, and apply this to get solutions of $7x + 8y + 9z = 1$.

66. In a rectangular coordinate system, put a dot at the point (n, m) if and only if n and m are relatively prime. Consider one-by-one boxes.
 (a) Can any such box have all four corners dotted?
 (b) Find a box with no corners dotted.
 (c) Let p be an odd prime. In the row of boxes

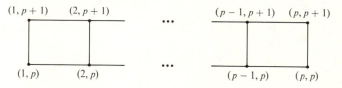

how many have three corners dotted?

67. Let $\langle x \rangle$ denote the fractional part of x. That is, $\langle x \rangle = x - [x]$.
 (a) Show that if $(a, n) = 1$, then the set of numbers

$$\langle a/n \rangle, \langle 2a/n \rangle, \ldots, \langle (n-1)a/n \rangle$$

is a permutation of the set

$$1/n, 2/n, \ldots, (n-1)/n.$$

 (b) Show that if $(a, n) = 1$, then

$$\sum_{k=0}^{n-1} \left[\frac{ak}{n} \right] = \frac{(a-1)(n-1)}{2}.$$

68. Show that every $n > 0$ satisfies at least one of

$$n \equiv 0 \ (\text{mod } 2) \qquad n \equiv 0 \ (\text{mod } 3) \qquad n \equiv 1 \ (\text{mod } 4)$$
$$n \equiv 3 \ (\text{mod } 8) \qquad n \equiv 7 \ (\text{mod } 12) \qquad n \equiv 23 \ (\text{mod } 24).$$

† 69. Show that 9^{9^9} ends in 89.

70. If p is a prime and $ap + b = c^2$, then show that all values of k making $kp + b$ a square are given by

$$k = pn^2 \pm 2cn + a,$$

where n is any integer.

† 71. If $m > 1$ is odd, show that $2^m + 1$ is composite.

72. Prove by induction that $3^{n+2} \mid 10^{3n} - 1$, $n = 0, 1, 2, \ldots$.

*† 73.
$$2^2(2^3 - 1) = 1^3 + 3^3;$$
$$2^4(2^5 - 1) = 1^3 + 3^3 + 5^3 + 7^3;$$
$$2^6(2^7 - 1) = 1^3 + 3^3 + \cdots + 15^3.$$

We might induce that every even perfect number, except 6, is a sum of consecutive odd cubes, starting with 1^3. Is this so?

74. (a) Given $n > 0$, show that there is an integer m such that

$$((n + 1)^{1/2} + n^{1/2})^2 = (m + 1)^{1/2} + m^{1/2}.$$

(b) Can you always find an m such that

$$((n + 1)^{1/2} + n^{1/2})^3 = (m + 1)^{1/2} + m^{1/2}?$$

† 75. Show that if n is an even perfect number, then $\prod_{d \mid n} d$ is a power of n.

76. If n is composite and greater than 4, show that $(n - 2)! \equiv 0 \pmod{n}$.

† 77. Show that the last nonzero digit of $n!$ is even when $n > 1$.

78. Show that no power of 2 is a sum of two or more consecutive positive integers.

† 79. Let $f(n)$ denote the smallest positive integer m such that $m! \equiv 0 \pmod{n}$.
(a) Make a table of f for $n = 2, 3, \ldots, 20$.
(b) Show that $f(p) = p$.
(c) Show that if p and q are distinct primes, then $f(pq) = \max(p, q)$.
(d) Show that if $p > k$, then $f(p^k) = kp$.

80. Show that

$$\sum_{k=1}^{n} k!$$

is never a square when $n > 3$.

* 81. Find nine integers in arithmetic progression whose sum of squares is a square.

82. Let p_i denote the ith prime. Show that $P_n = p_1 p_2 \cdots p_n + 1$ is never a square.

*† 83. Which positive integers are neither composite nor the sum of two positive composite integers?

84. $1000 \equiv 1 \pmod{37}$. From this, develop a test for the divisibility of an integer by 37.

*† 85. Although 9 is not the sum of two positive integer squares, it is the sum of two positive *rational* squares. Find them.

86. Suppose that we have a solution of

$$ab(a + b)(a - b) = c^2$$

where a, b, and c have no common factors.

(a) Show that a and b are both odd.

(b) Show that any two of a, b, $(a + b)/2$, and $(a - b)/2$ are relatively prime.

(c) Conclude that each of a, b, $(a + b)/2$, and $(a - b)/2$ is a square.

(d) Put $(a + b)/2 = r^2$ and $(a - b)/2 = s^2$. What are a and b in terms of r and s?

(e) Conclude that if there is a solution of $ab(a^2 - b^2) = c^2$ where a, b, c have no common factor, then there is a solution of

$$r^2 + s^2 = t^2,$$
$$r^2 - s^2 = u^2.$$

87. Euler showed that any odd perfect number must be of the form

$$p^{4a+1}Q^2,$$

p an odd prime, a and Q integers. Fill in any missing details in this sketch of his proof:

Let $n = P_1 P_2 \cdots P_k$ be the decomposition of n into powers of distinct odd primes. Let $Q_i = \sigma(P_i)$, $i = 1, 2, \ldots, k$. If $\sigma(n) = 2n$, then

$$2P_1 P_2 \cdots P_k = Q_1 Q_2 \cdots Q_k.$$

Thus, one of Q_1, Q_2, \ldots, Q_k—say Q_1—is double an odd number, and the remaining ones are odd. Thus $P_2, P_3, \ldots P_k$ are even powers of primes. Also, $P_1 = p^{4a+1}$ for some prime p and integer a.

88. Write the positive integers in a spiral-like array as shown:

17	16	15	14	13
18	5	4	3	12
19	6	1	2	11
20	7	8	9	10
21	22	23	. . .	

If 1 is at the origin of a rectangular coordinate system and $n \geq 0$,

(a) What integer is at $(n, 0)$?

(b) What integer is at (n, n)?

(c) What integer is at $(-n, 0)$?

(d) Where is $(2n + 1)^2$?

(e) Where is $(2n)^2$?

(f) Where is 1000?

89. Show that the primes less than n^2 are the odd numbers not included in the arithmetic progressions

$$r^2, r^2 + 2r, r^2 + 4r, \ldots \text{ (up to } n^2)$$

for $r = 3, 5, 7, \ldots$ (up to $n - 1$).

90. (a) Find all x, y, z in arithmetic progression such that

$$x^2 + xy + y^2 = z^2$$

and $xy \neq 0$.

(b) Find all x, y, z in arithmetic progression such that

$$x^2 + kxy + y^2 = z^2$$

and $xy \neq 0$.

† 91. Show that $x^n + y^n = z^n$ has no solutions with both x and y less than n for any positive integer n.

92. Suppose that $n^2 + n + 1 = p^r$, where $n \geq 1$, p is prime, and $r \geq 1$.
(a) Show that p is odd.
(b) Show that if $n \equiv 1 \pmod 3$, then the only solution is $n = 1$, $p = 3$, $r = 1$.
(c) Show that r is odd.
(d) Show that if $p \neq 3$, then $p \equiv 1 \pmod 3$.

† 93. Let $f(x) = a_n x^n + a_{n-1} x^{n-1} + \cdots + a_0$. Suppose that a_n, a_0, and an odd number of the remaining coefficients are odd. Show that $f(x) = 0$ has no rational roots.

94. (a) Show that if $(a, p) = 1$, $n \mid (p - 1)$, and $a^{(p-1)/n} \not\equiv 1 \pmod p$, then a is not an nth power residue $\pmod p$.
(b) Is 2 a fifth power residue $\pmod{31}$?

† 95. Show that $n \mid (2^n + 1)$ if n is a power of 3.

96. (a) Show that $n^2 + (n + 1)^2 = 3m^2$ is impossible.
(b) What is a sufficient condition for

$$n^2 + (n + 1)^2 = km^2$$

to have a solution for given k, $k > 0$?

*† 97. Let m be square-free. (That is, $m = p_1 p_2 \cdots p_k$, a product of *distinct* primes.) Suppose that m has the property that $p \mid m$ implies $(p - 1) \mid m$.
(a) Show that $m = 2$, 6, and 42 have this property.
(b) Does any other m?
(c) How many others?

98. (a) Show that if r and s satisfy $5^n s - 2^n r = 1$ and $x = 5^n s$, then the last n digits of x^2 are the same as the last n digits of x.
(b) Find such a number for $n = 3$.

† 99. Show that the sequence $\{2 + np\}$, p an odd prime, $n = 1, 2, \ldots$, contains an infinite geometric progression for any p.

100. Solve $x(x - 31) = y(y - 41)$ in positive integers.

Proof by Induction

In the text, the method of proof by mathematical induction is used several times. The purpose of this section is to recall what the method is, show some examples of how it operates, and give some problems for practice.

Mathematics is notoriously a deductive art: starting with a collection of postulates, theorems are deduced by following the laws of logic. That is the way it is presented in print, but that is not the way that new mathematics is discovered. It is difficult to sit down and think, "I will now deduce," and deduce anything worthwhile. The goal must be in sight: you must suspect that a theorem is true, and then deduce it from what you know. The theorem you suspect is true must come from somewhere. Just as in the other sciences, a new result can come at any time, the mysterious product of inspiration, inspection, subconscious rumination, revelation, or a correct guess.

[*] **Exercise 1.** Guess what $f(n)$ is from the following data:

n	0	1	2	3	4	5
$f(n)$	1	0	1	4	9	16

* **Exercise 2.** Guess what $f(n)$ is from

n	0	1	2	3	4	5	6
$f(n)$	1	2	5	10	17	26	37

* **Exercise 3** (optional). Guess a theorem about $f(n)$:

n	1	2	3	4	5	6	7
$f(n)$	2	-1	-2	-1	2	7	14

Since number theory is largely concerned with the positive integers, some of its theorems are of the form, "Such-and-such is true for all positive integers n." Propositions like this can often be proved by *mathematical induction* (or *induction* for short—we will not be concerned with any other kind). This method of proof is based on the following property of positive integers:

(1) If a set of integers contains 1, and
 if it contains $r + 1$ whenever it contains r,
 then the set contains all the positive integers.

This property is so fundamental that it is usually taken as a nonprovable postulate about the positive integers. It is applied when we want to show that a proposition $P(n)$ about the positive integer n is true for all n, $n = 1, 2, \ldots$. Examples of such propositions are

$P_1(n)$: "$n^2 + 3n + 2 > (n + 1)^2 - 5$."
$P_2(n)$: "$n(n + 1)(n + 2)$ is divisible by 6."
$P_3(n)$: "$f(x_1 + x_2 + \cdots + x_n) \geq f(x_1) + f(x_2) + \cdots + f(x_n)$."

Let S denote the set of positive integers for which $P(n)$ is true. If we can show that 1 is in S and that if r is in S, then $r + 1$ is in S, then (1) says that all positive integers are in S. Rephrasing this, we get the induction principle:

If $P(1)$ is true, and
if the truth of $P(r)$ implies the truth of $P(r + 1)$,
then $P(n)$ is true for all n, $n = 1, 2, \ldots$.

There is no necessity for an induction to start with 1; we could start with 2, 3, 17, or -12. For example, if $P(3)$ is true and if the truth of $P(r)$ implies the truth of $P(r + 1)$, then $P(n)$ is true for $n = 3, 4, \ldots$.

* **Exercise 4.** Fill in the blank: if $P(2)$ is true, and if the truth of $P(r)$ implies the truth of $P(r + 2)$, then $P(n)$ is true for _____ .

To illustrate a proof by induction, we will take a well known example. Let $P(n)$ be the statement

$$\text{``}1 + 2 + \cdots + n = n(n + 1)/2.\text{''}$$

We will prove by induction that $P(n)$ is true for all positive integers n.

* **Exercise 5.** What is $P(1)$? Is it true?

Suppose that $P(r)$ is true; that is, suppose that

(2) $1 + 2 + \cdots + r = r(r + 1)/2.$

We wish to deduce that $P(r + 1)$ is true. That is, we want to show that

(3) $1 + 2 + \cdots + (r + 1) = (r + 1)(r + 2)/2$

follows from (2). If we add $r + 1$ to both sides of (2), we get

$$1 + 2 + \cdots + r + (r + 1) = r(r + 1)/2 + (r + 1)$$

$$= (r + 1)\left(\frac{r}{2} + 1\right) = (r + 1)(r + 2)/2,$$

which is (3). Hence both parts of the induction principle have been verified, and it follows that $P(n)$ is true for all positive integers n.

In any proof by induction, we must not forget to show that $P(1)$ is true. Even if we show that the truth of $P(r)$ implies the truth of $P(r + 1)$, if $P(1)$ is not true, then we cannot conclude that $P(n)$ is true for any n. For example, let $P(n)$ be

$$n + (n + 1) = 2n.$$

Suppose that $P(r)$ is true. That is, we assume that

(4) $r + (r + 1) = 2r.$

Using this, we have

$$(r + 1) + (r + 2) = r + (r + 1) + 2 = 2r + 2 = 2(r + 1),$$

so $P(r + 1)$ is true. So, *if* $P(1)$ were true, it would follow that $P(n)$ is true for all positive integers n. Since $P(1)$ is not true, we cannot so conclude. In fact, $P(n)$ is false for all n.

It should go without saying that in any proof by induction, we must verify that the truth of $P(r)$ implies the truth of $P(r + 1)$. For example, from the table

n	1	2	3	4	5	6
$f(n)$	2	4	6	8	10	12

we cannot conclude that $f(n) = 2n$ for all n. In fact, $f(7) = \pi$, because the function that I had in mind when constructing the table was

$$f(n) = 2n + \frac{(n-1)(n-2)(n-3)(n-4)(n-5)(n-6)(\pi - 14)}{6 \cdot 5 \cdot 4 \cdot 3 \cdot 2}.$$

Another form of the induction principle is sometimes used:

If $P(1)$ is true, and
if the truth of $P(k)$ for $1 \le k \le r$ implies the truth of $P(r + 1)$,
then $P(n)$ is true for all n, $n = 1, 2, \ldots$.

This is valid because of the corresponding property of integers: if a set of integers contains 1, and contains $r + 1$ whenever it contains $1, 2, \ldots, r$, then it contains all positive integers.

Problems

1. Prove that

$$1^2 + 2^2 + \cdots + n^2 = n(2n + 1)(n + 1)/6$$

for $n = 1, 2, \ldots$.

2.
$$1^3 = 1^2,$$
$$1^3 + 2^3 = 3^2,$$
$$1^3 + 2^3 + 3^3 = 6^2,$$
$$1^3 + 2^3 + 3^3 + 4^3 = 10^2.$$

Guess a theorem and prove it.

† 3. From Problem 2, or by guessing and induction, derive a formula for

$$1^3 + 3^3 + 5^3 + \cdots + (2k - 1)^3,$$

$k = 1, 2, \ldots$.

4. Prove that

$$1/1 \cdot 2 + 1/2 \cdot 3 + \cdots + 1/(n - 1)n = 1 - (1/n)$$

for $n = 2, 3, \ldots$.

† 5. 1, 3, 6, and 10 are called **triangular numbers**:

Let t_n denote the nth triangular number. Find a formula for t_n.

6. Suppose that $a_1 = 1$ and $a_{n+1} = 2a_n + 1$, $n = 1, 2, \ldots$. Prove by induction that $a_n = 2^n - 1$.

7. Suppose that $a_0 = a_1 = 1$ and $a_{n+1} = a_n + 2a_{n-1}$, $n = 1, 2, \ldots$ Prove by induction that

$$a_n = \frac{2^{n+1} + (-1)^n}{3}.$$

8. Suppose that $a_1 = a_2 = 1$ and $a_{n+1} = 3a_n + a_{n-1}$. Prove that $(a_n, a_{n+1}) = 1$, $n = 1, 2, \ldots$.

† 9.

$$1 \cdot 2 \cdot 3 \cdot 4 = 5^2 - 1,$$
$$2 \cdot 3 \cdot 4 \cdot 5 = 11^2 - 1,$$
$$3 \cdot 4 \cdot 5 \cdot 6 = 19^2 - 1,$$
$$4 \cdot 5 \cdot 6 \cdot 7 = 29^2 - 1.$$

Guess and prove a theorem. (Induction may not be necessary.)

10. Guess and prove a formula for

$$1^2 + 4^2 + 7^2 + \cdots + (3n + 1)^2,$$

$n = 0, 1, \ldots$.

11. Prove by induction that $n(n + 1)(n + 2)$ is divisible by 6 for $n = 1, 2, \ldots$.

12. Construct a formula for a function f such that

$$f(1) = f(2) = f(3) = f(4) = 0, \qquad f(5) = 17.$$

*† 13. Let t_n denote the nth triangular number. Consider the table

n	1	2	3	4	5
t_n	1	3	6	10	15
$8t_n + 1$	9	25	49	81	121.

Are all those squares a coincidence?

14. Prove by induction that $n^5 - n$ is divisible by 5, $n = 1, 2, \ldots$.

† 15. The **Fibonacci numbers** are defined by

$$f_{n+1} = f_n + f_{n-1}, \qquad f_1 = f_2 = 1.$$

Prove that f_{5n} is divisible by 5, $n = 1, 2, \ldots$.

Appendix

Computer Problems

Computers go well with number theory. The reason is that what computers do best is long sequences of calculations, with a single number as a result, and that is often just what is wanted to solve a problem. It would be hard to solve a congruence like $314159x \equiv 26535$ (mod 27182818) by hand, but quite easy to program a computer—or even a programmable pocket calculator—to find the solutions: have the machine do the labor of substituting in $x = 1, 2, \ldots, 27182818$ and check if the congruence is satisfied. Even if the machine is so slow that it can do only 1000 substitutions and checks per second, it will take no more than eight hours for it to do them all. Plug it in and let it run all night: that is what computers like to do. It is the same when looking for solutions to a diophantine equation or seeing if a large integer is prime: the machine can grind through hundreds of millions of calculations without complaint and produce results which human life would be too short to obtain otherwise.

Applying computers to number theory can also sharpen programming skills. Anyone can write a satisfactory program to print paychecks, since most of the time is spent by the relatively slow printer. But when a problem requires all the computing power of a machine, the algorithm being used can make quite a difference. I have seen programs that

210

factor nine-digit integers almost instantly and others that take months. When the machine is being used to its utmost, there is pressure on the programmer to push himself to his limits, and this can have only good effects.

Some things which could be programmed are obvious, but nevertheless are listed below.

Write a program to:

Find greatest common divisors using the Euclidean Algorithm.

Solve $ax + by = (a, b)$.

Determine if an integer is prime.

Factor integers (a good factoring subroutine is very helpful in many problems).

Solve $ax + by = c$.

Solve simultaneous congruences using the Chinese Remainder Theorem.

Solve $ax \equiv b \pmod{m}$.

Calculate the least residue of $a^k \pmod{m}$.

Calculate $d(n)$, $\sigma(n)$, $\phi(n)$.

Determine if an integer is deficient or abundant.

Find primitive roots of an integer.

Find for which integers a given number is a primitive root.

Solve $x^2 \equiv a \pmod{m}$.

Solve $ax^2 + bx + c \equiv 0 \pmod{m}$.

Evaluate (a/p).

Find the representation of an integer in one base, given its representation in another.

Find the period of the decimal expansion of $1/n$.

Find representations of integers as sums of squares, cubes, or higher powers.

Find solutions of $x^2 - Ny^2 = 1$, or $x^2 - Ny^2 = k$.

Evaluate $\pi(x)$.

There follow 25 other possibilities, listed essentially at random. I do not know how hard or easy, or how frustrating or rewarding, each one is. They are examples of things that have been found interesting, and some references have been included.

1. An amusing game, a bit like writing integers using exactly four 4s $(1 = 44/44, \ 2 = 4/4 + 4/4, \ 3 = (4 + 4 + 4)/4, \ . . .)$, is to write each prime as a difference of two integers whose prime-power decompositions include just the smaller primes. For example, $5 = 3^2 - 2^2$ and $7 = 5^2 - 2 \cdot 3^2$. The game gets so hard for larger primes—29 is $3 \cdot 11 \cdot 13^2 \cdot 19 \cdot 23 - 2^{12} \cdot 5 \cdot 7 \cdot 17$—that machine help is needed. This may not be easy to program. (*Mathematical Reviews* **49**(1975):#4921.)

2. The answer to this question is known, but it illustrates how the computer can be used to gather data from which inductions can be made: What is $\Sigma(a - 1, n)$ in terms of n, where the sum is taken over those integers from 1 to n which are relatively prime to n? (*Mathematical Reviews* **49**(1975):#2506.)

3. It is striking that $145 = 1! + 4! + 5!$, and the only other integers satisfying $d_1 d_2 \cdots d_k = \sum_{i=1}^{k} d_i!$ are 1, 2, and 40585 (*Mathematics Magazine* **44**(1971):278–279). Does $d_1 d_2 \cdots d_k$ ever equal $\sum_{i=1}^{k} 2^{d_i}$ or $\sum_{i=1}^{k} 3^{d_i}$?

4. If $n = 6$, or if $n \equiv 1 \pmod 6$ is prime, or if $n = 3p$ where $p \equiv 5 \pmod 6$, then $3 \mid n + \sigma(n)$. Are there any other such n, and do they fall into classes?

5. The equation $\sigma(n) = \sigma(n + 1)$ has 113 solutions for $n < 10^7$ (*Mathematics of Computation* **27**(1973):676). It is probably too hard to extend that table, but related equations like $\sigma(n + 1) = 2\sigma(n)$ or $\sigma(n + 2) = \sigma(n)$ might yield interesting numbers.

6. Let $s(n) = \sigma(n) - n$, $s^2(n) = s(s(n))$, and $s^{k+1}(n) = s(s^k(n))$. Catalan's Conjecture, which dates back to 1887, is that the sequence $n, s(n)$, $s^2(n), \ . . .$ eventually either reaches 1 or enters a cycle. The general opinion now is that the conjecture is false and that for many n the sequence is unbounded. A computer will never be able to establish that, of course, but it can discover cycles. Fourteen cycles of period 4 exist (for example, $s^5(12496) = 12496$) and there are longer cycles ($s^{28}(14316) = 14316$). A machine might discover more. It is likely that they would be rediscoveries though, so a variation of Catalan's Conjecture might be considered, say by defining $t(n) = \sigma(n) - n - 1$.

7. An integer n is **semiperfect** if n is a sum of distinct proper divisors of n. For example, $104 = 2^3 \cdot 13$ is semiperfect because $104 = 52 + 26 + 13 + 8 + 4 + 1$. A semiperfect integer is **irreducible** if it is not the multiple of a smaller semiperfect number. For example, 104 is irreducibly semiperfect, since none of 2, 4, 8, 13, 26, or 52 is semiper-

fect. There is a theorem that $n = 2^m p$ is irreducibly semiperfect if $2^m < p < 2^{m+1}$. Are there other irreducible semiperfect numbers? If there are, does some theorem exist to explain them? (*Mathematical Reviews* **50**(1975): #12905).

8. The equation $1 + 2 + \cdots + n = 1^2 + 2^2 + \cdots + k^2$ has just four solutions, three being (1, 1), (10, 5), and (13, 6). Find the fourth (n is less than 1000), and if a program can do that quickly, adapt it to consider related equations like $1^r + 2^r + \cdots + n^r = 1^s + 2^s + \cdots + k^s$ or $1 + 2 + \cdots + n = a^2 + (a + 1)^2 + \cdots + k^2$ (*Mathematical Reviews* **46**(1973): #3443, #8967).

9. Ever since the Egyptians were building pyramids, people have been concerned with how to write fractions as sums of reciprocals. It is known that a/n is a sum of three reciprocals, given a, for all sufficiently large n. When $a = 4$, such representations are known for all $n < 10^7$, and when $a = 5$ for $n < 922321$. Tables of solutions for those values, or for $a = 6, 7, \ldots$ might disclose interesting patterns. And they might not, but one never knows until one tries (*Mathematical Reviews* **44**(1972): #6600; *Mathematics Magazine* **46**(1973):241–244).

10. Given sets of integers $A = \{a_1, a_2, \ldots, a_n\}$ and $B = \{b_1, b_2, \ldots, b_k\}$, define $A + B$ to be $\{a_i + b_j\}$ and $A - B$ to be $\{a_i - b_j\}$, where $i = 1, 2, \ldots, n$ and $j = 1, 2, \ldots, k$. For example, if $A = \{1, 2, 3\}$, then $A + A = \{2, 3, 4, 5, 6\}$ and $A - A = \{-2, -1, 0, 1, 2\}$. J. H. Conway conjectured that the number of elements in $A - A$ is never less than the number of elements in $A + A$. This is false (*Mathematical Reviews* **40**(1970): #2635). One counterexample is $\{1, 2, 3, 5, 8, 9, 13, 15, 16\}$. Other counterexamples could be searched for, but it might be more interesting to investigate sets like $(A + A) - A$ and $A + (A - A)$.

11. Let $f(n) = n - \phi(n)$, $f^2(n) = f(f(n))$, and $f^{k+1}(n) = f(f^k(n))$, and consider the sequence $n, f(n), f^2(n), \ldots$. Since $f(n) < n$, it always reaches 1. For example, if $n = 100$, the sequence is 60, 44, 24, 16, 8, 4, 2, 1: eight steps are needed. Let $s(k)$ be the smallest integer which reaches 1 in k steps. The first few values of s are

k	2	3	4	5	6	7
$s(k)$	4	6	10	18	30	42

and when factored, they are $2 \cdot 2$, $2 \cdot 3$, $2 \cdot 5$, $2 \cdot 3 \cdot 3$, $2 \cdot 3 \cdot 5$, and $2 \cdot 3 \cdot 7$. Could anyone have so little curiosity as to not want to know if $s(8)$ was $2 \cdot 3 \cdot 5 \cdot 7$, $2 \cdot 3 \cdot 3 \cdot 5$, or $2 \cdot 3 \cdot 11$? These sequences, analogous to the sequences in Catalan's Conjecture, have yet to be investi-

gated, I believe, though the sequences n, $\phi(n)$, $\phi(\phi(n))$, . . . have been studied by many writers.

12. Generalize Wilson's Theorem by looking at the values of $(2p - 1)!$ (mod p^2), $(3p - 1)!$ (mod p^3), $(4p - 1)!$ (mod p^4), The result is known but not easy to prove: with the aid of a machine it could be rediscovered (*Mathematical Reviews* **42**(1971):#4477). One might then generalize the theorem in a different way, say by looking at $a(a + 1)$. . . $(a + p)$ (mod p).

13. Let $s(n) = \sum_{r=1}^{n} (r, n)$. The first few values of this function, first studied by S. S. Pillai in 1933, are 1, 3, 5, 8, 9, 15, 13, 20, 21, and 25. Properties of this sequence might be investigated. It surely has some, since $s(n) - \sigma(n)$ is 0, 0, 1, 1, 3, 3, 5, 5, 8, 8 for $n = 1, 2, . . . , 10$, and can that be coincidence? (*Mathematical Reviews* **48**(1974): #2508).

14. A search in 1972 for solutions of $a(x^n - 1)/(x - 1) = y^m$ with $1 < a < x \le 10$, $n > 2$, and $m \ge 2$ yielded only $n = a = 4$, $x = 7$, $m = 2$, and $y = 40$ (*Mathematical Reviews* **46**(1973): #1703). It is quite possible that there are no others, but it might be fun to put $a = 11$ and start looking.

15. Triangular numbers (1, 3, 6, 10, . . .) have the form $n(n + 1)/2$. Palindromes (11, 232, 36763) read the same backward as forward. Find some palindromic triangular numbers (*Fibonacci Quarterly* **12**(1974): 209–212), or, if you do not like to repeat work already done, find palindromic pentagonal numbers.

16. Perfect numbers (those such that $\sigma(n) = 2n$) and superperfect numbers (those such that $\sigma(\sigma(n)) = 2n$) have been investigated. But no one has looked into superduperperfect numbers (those such that $\sigma(\sigma(\sigma(n))) = 2n$), and there may not even be any. With a good subroutine for evaluating $\sigma(n)$, it shouldn't be hard to find any that exist, and if there are none, to find solutions of variants like $\sigma(\sigma(\sigma(n))) = 3n$ or $\sigma(\sigma(\sigma(\sigma(n)))) = 16n$.

17. The congruences $n\sigma(n) \equiv 2$ (mod $\phi(n)$) and $\phi(n) d(n) \equiv -2$ (mod n) are both true if n is a prime. Find composite solutions. The answers are known (*Mathematical Reviews* **50**(1975):#2049), but there are other such congruences, like $\phi(n)\sigma(n) \equiv -1$ (mod n^2), that no one has looked at.

18. If $n = p_1^{e_1} p_2^{e_2} \cdots p_k^{e_k}$, define f by $f(1) = 2$ and $f(n) = 1 + e_1 p_1 + e_2 p_2 + \cdots + e_k p_k$. See what happens to the sequences n, $f(n)$, $f(f(n))$,

. . . for various n. The answer here is also known (*Mathematical Reviews* **50**(1975): #220), but letting $g(n) = e_1 p_1 + e_2 p_2 + \cdots + e_k p_k$ or $1 + 2(e_1 p_1 + e_2 p_2 + \cdots + e_k p_k)$ would lead to sequences that might be dramatically different.

19. All of the integers n such that $\phi(n) = \phi(n + 1)$ are known up to 2792144 (*Mathematics of Computation* **29**(1975):321), and there is only one integer in that range such that $\phi(n) = \phi(n + 1) = \phi(n + 2)$. To extend the table of solutions beyond 2792144 would take a powerful machine, but equations like $\phi(n) = \phi(n + 2)$ are mostly uninvestigated, and triplets such that $\phi(n) = \phi(n + 2) = \phi(n + 4)$ or $\phi(n) = \phi(n + 3) = \phi(n + 6)$ would be picturesque.

20. There are many more n such that $d(n) = d(n + 1)$, and in fact triplets are common, the first being $d(33) = d(34) = d(35) = 4$. It would be interesting to know where the first quadruplet, quintuplet, . . . were.

21. Every integer greater than 20161 is a sum of two abundant numbers [13]. What is the largest integer that is not the sum of 3, 4, 5, . . . abundant numbers?

22. Given n, let a_r be its least residue (mod r) and let $s(n) = a_1 + a_2 + \cdots + a_n$. The first few values of $s(n)$ are

n	1	2	3	4	5	6	7	8	9	10	11	12	13
$s(n)$	0	0	1	1	4	5	8	8	12	13	22	17	28

which looks irregular, but there are regularities: for example, it has been known for a long time that $s(n) - s(n - 1) + \sigma(n) = 2n - 1$ (*Mathematical Reviews* **50**(1975):#7014). It would be interesting to have data about solutions of equations like $s(n) = n$, $s(n) = 2n$, $s(n) = n/2$, and so on; some discovery might be waiting to be made.

23. Let $L(n) = \sigma(n) - (m + 1)$; $L(n)$ is the sum of the proper divisors of n. If $L(m) = n$ and $L(n) = m$, then m and n are a *reduced amicable pair*. They have been tabulated up to 100000 (*Mathematics of Computation* **25**(1971):923–925). It would be possible to carry the tabulation further, or to investigate variations. For example, if n is even, let $K(n)$ be the sum of the *unobvious proper divisors* of n. That is, $K(n) = \sigma(n) - (n + n/2 + 1)$. Continuing in a parody of high academic style, define *severely reduced amicable pairs* as those integers m and n such that $K(m) = n$ and $K(n) = m$. Then go to it: a whole new field lies unexplored before you.

24. Carmichael's Conjecture that $\phi(x) = n$ never has exactly one so-

lution if $n \geq 2$ will probably not be settled soon. If such an n exists, it is greater than 10^{400} (*Mathematical Reviews* **49**(1975): #4917), and those numbers are too large for even the most powerful computer to handle. For any n, there are only a finite number of solutions, and a table exists (*Mathematical Reviews* **41**(1971): #5291) giving them up to $n = 1978$. A large enough machine could extend that table, and even a small one could compile a table of solutions of $\sigma(x) = n$, which also has only finitely many solutions for given n.

25. Every positive integer is a sum of four squares and of nine cubes. What if squares and cubes are mixed? It is probably false that every positive integer is a sum of a cube and three squares, but how many squares are needed so that every integer is a sum of two cubes and some squares? It is true that every positive integer is a sum of eight cubes and a square, but can it be done with seven cubes and a square? Although a computer cannot prove any theorems, it can indicate what theorems to try to prove.

Appendix

C

Factor Table for Integers Less Than 10,000

The following table gives the smallest prime factor of each odd positive integer n, $3 \le n \le 9999$, not divisible by five. The numbers across the top of each column—1, 3, 7, 9—give the units' digit of n and the numbers down the side give the thousands', hundreds', and tens' digits of n. A dash in the table indicates that n is prime.

For example, reading across line 40 of the table, we see that 401 and 409 are prime, 403 is divisible by 13, and 407 is divisible by 11. With the aid of the table, the prime-power decomposition of any integer less than 10,000 (and of any even integer less than 20,000) can be quickly determined. For example, take 759. From line 75 of the table, we see that $3 \mid 759$, and a division shows that $759 = 3 \cdot 253$. From line 25 of the table, $11 \mid 253$, so $759 = 3 \cdot 11 \cdot 23$. Line 2 of the table shows that 23 is prime, so we have the prime-power decomposition of 759.

	1 3 7 9		1 3 7 9		1 3 7 9		1 3 7 9
0	—— 3	**40**	— 13 11 —	**80**	3 11 3 —	**120**	— 3 17 3
1	————	41	3 7 3 —	81	— 3 19 3	121	7 ——— 23
2	3 — 3 —	42	— 3 7 3	82	————	122	3 — 3 —
3	— 3 — 3	43	—— 19 —	83	3 7 3 —	123	— 3 — 3
4	——— 7	44	3 — 3 —	84	29 3 7 3	124	17 11 29 —
5	3 — 3 —	45	11 3 — 3	85	23 ———	125	3 7 3 —
6	— 3 — 3	46	——— 7	86	3 — 3 11	126	13 3 7 3
7	—— 7 —	47	3 11 3 —	87	13 3 — 3	127	31 19 ———
8	3 — 3 —	48	13 3 — 3	88	——— 7	128	3 — 3 —
9	7 3 — 3	49	— 17 7 —	89	3 19 3 29	129	— 3 — 3
10	————	**50**	3 — 3 —	**90**	17 3 — 3	**130**	——— 7
11	3 — 3 7	51	7 3 11 3	91	— 11 7 —	131	3 13 3 —
12	11 3 — 3	52	—— 17 23	92	3 13 3 —	132	— 3 — 3
13	— 7 ——	53	3 13 3 7	93	7 3 — 3	133	11 31 7 13
14	3 11 3 —	54	— 3 — 3	94	— 23 — 13	134	3 17 3 19
15	— 3 — 3	55	19 7 — 13	95	3 8 3 7	135	7 3 23 3
16	7 —— 13	56	3 — 3 —	96	31 3 — 3	136	— 29 — 37
17	3 — 3 —	57	— 3 — 3	97	— 7 — 11	137	3 — 3 7
18	— 3 11 3	58	7 11 — 19	98	3 — 3 23	138	— 3 19 3
19	————	59	3 — 3 —	99	— 3 — 3	139	13 7 11 —
20	3 7 3 11	**60**	— 3 — 3	**100**	7 17 19 —	**140**	3 23 3 —
21	— 3 7 3	61	13 ———	101	3 — 3 —	141	17 3 13 3
22	13 ———	62	3 7 3 17	102	— 3 13 3	142	7 ———
23	3 — 3 —	63	— 3 7 3	103	—— 17 —	143	3 — 3 —
24	— 3 13 3	64	——— 11	104	3 7 3 —	144	11 3 — 3
25	— 11 — 7	65	3 — 3 —	105	— 3 7 3	145	—— 31 —
26	3 — 3 —	66	— 3 23 3	106	—— 11 —	146	3 7 3 13
27	— 3 — 3	67	11 ——— 7	107	3 29 3 13	147	— 3 7 3
28	—— 7 17	68	3 — 3 13	108	23 3 — 3	148	————
29	3 — 3 13	69	— 3 17 3	109	——— 7	149	3 — 3 —
30	7 3 — 3	**70**	— 19 7 —	**110**	3 — 3 —	**150**	19 3 11 3
31	——— 11	71	3 23 3 —	111	11 3 — 3	151	— 17 37 7
32	3 17 3 7	72	7 3 — 3	112	19 — 7 —	152	3 — 3 11
33	— 3 — 3	73	17 — 11 —	113	3 11 3 17	153	— 3 29 3
34	11 7 ——	74	3 — 3 7	114	7 3 31 3	154	23 — 7 —
35	3 — 3 —	75	— 3 — 3	115	—— 13 19	155	3 — 3 —
36	19 3 — 3	76	— 7 13 —	116	3 — 3 7	156	7 3 — 3
37	7 — 13 —	77	3 — 3 19	117	— 3 11 3	157	— 11 19 —
38	3 — 3 —	78	11 3 — 3	118	— 7 — 29	158	3 — 3 7
39	17 3 — 3	79	7 13 — 17	119	3 — 3 11	159	37 3 — 3

	1	3	7	9		1	3	7	9		1	3	7	9		1	3	7	9
160	—	7	—	—	**200**	3	—	3	7	**240**	7	3	29	3	**280**	—	—	7	53
161	3	—	3	—	201	—	3	—	3	241	—	19	—	41	281	3	29	3	—
162	—	3	—	3	202	43	7	—	—	242	3	—	3	7	282	7	3	11	3
163	7	23	—	11	203	3	19	3	—	243	11	3	—	3	283	19	—	—	17
164	3	31	3	17	204	13	3	23	3	244	—	7	—	31	284	3	—	3	7
165	13	3	—	3	205	7	—	11	29	245	3	11	3	—	285	—	3	—	3
166	11	—	—	—	206	3	—	3	—	246	23	3	—	3	286	—	7	47	19
167	3	7	3	23	207	19	3	31	3	247	7	—	—	37	287	3	13	3	—
168	41	3	7	3	208	—	—	—	—	248	3	13	3	19	288	43	3	—	3
169	19	—	—	—	209	3	7	3	—	249	47	3	11	3	289	7	11	—	13
170	3	13	3	—	**210**	11	3	7	3	**250**	41	—	23	13	**290**	3	—	3	—
171	29	3	17	3	211	—	—	29	13	251	3	7	3	11	291	41	3	—	3
172	—	—	11	7	212	3	11	3	—	252	—	3	7	3	292	23	37	—	29
173	3	—	3	37	213	—	3	—	3	253	—	17	43	—	293	3	7	3	—
174	—	3	—	3	214	—	—	19	7	254	3	—	3	—	294	17	3	7	3
175	17	—	7	—	215	3	—	3	17	255	—	3	—	3	295	13	—	—	11
176	3	41	3	29	216	—	3	11	3	256	13	11	17	7	296	3	—	3	—
177	7	3	—	3	217	13	41	7	—	257	3	31	3	—	297	—	3	13	3
178	13	—	—	—	218	3	37	3	11	258	29	3	13	3	298	11	19	29	7
179	3	11	3	7	219	7	3	13	3	259	—	—	7	23	299	3	41	3	—
180	—	3	13	3	**220**	31	—	—	47	**260**	3	19	3	—	**300**	—	3	31	3
181	—	7	23	17	221	3	—	3	7	261	7	3	—	3	301	—	23	7	—
182	3	—	3	31	222	—	3	17	3	262	—	43	37	11	302	3	—	3	13
183	—	3	11	3	223	23	7	—	—	263	3	—	3	7	303	7	3	—	3
184	7	19	—	43	224	3	—	3	13	264	19	3	—	3	304	—	17	11	—
185	3	17	3	11	225	—	3	37	3	265	11	7	—	—	305	3	43	3	7
186	—	3	—	3	226	7	31	—	—	266	3	—	3	17	306	—	3	—	3
187	—	—	—	—	227	3	—	3	43	267	—	3	—	3	307	37	7	17	—
188	3	7	3	—	228	—	3	—	3	268	7	—	—	—	308	3	—	3	—
189	31	3	7	3	229	29	—	—	11	269	3	—	3	—	309	11	3	19	3
190	—	11	—	23	**230**	3	7	3	—	**270**	37	3	—	3	**310**	7	29	13	—
191	3	—	3	19	231	—	3	7	3	271	—	—	11	—	311	3	11	3	—
192	17	3	41	3	232	11	23	13	17	272	3	7	3	—	312	—	3	53	3
193	—	—	13	7	233	3	—	3	—	273	—	3	7	3	313	31	13	—	43
194	3	29	3	—	234	—	3	—	3	274	—	13	41	—	314	3	7	3	47
195	—	3	19	3	235	—	13	—	7	275	3	—	3	31	315	23	3	7	3
196	37	13	7	11	236	3	17	3	23	276	11	3	—	3	316	29	—	—	—
197	3	—	3	—	237	—	3	—	3	277	17	47	—	7	317	3	19	3	11
198	7	3	—	3	238	—	—	7	—	278	3	11	3	—	318	—	3	—	3
199	11	—	—	—	239	3	—	3	—	279	—	3	—	3	319	—	31	23	7

	1	3	7	9		1	3	7	9		1	3	7	9		1	3	7	9
320	3	—	3	—	**360**	13	3	—	3	**400**	—	—	—	19	**440**	3	7	3	—
321	13	3	—	3	361	23	—	—	7	401	3	—	3	—	441	11	3	7	3
322	—	11	7	—	362	3	—	3	19	402	—	3	—	3	442	—	—	19	43
323	3	53	3	41	363	—	3	—	3	403	29	37	11	7	443	3	11	3	23
324	7	3	17	3	364	11	—	7	41	404	3	13	3	—	444	—	3	—	3
325	—	—	—	—	365	3	13	3	—	405	—	3	—	3	445	—	61	—	7
326	3	13	3	7	366	7	3	19	3	406	31	17	7	13	446	3	—	3	41
327	—	3	29	3	367	—	—	—	13	407	3	—	3	—	447	17	3	11	3
328	17	7	19	11	368	3	29	3	7	408	7	3	61	3	448	—	—	7	67
329	3	37	3	—	369	—	3	—	3	409	—	—	17	—	449	3	—	3	11
330	—	3	—	3	**370**	—	7	11	—	**410**	3	11	3	7	**450**	7	3	—	3
331	7	—	31	—	371	3	47	3	—	411	—	3	23	3	451	13	—	—	—
332	3	—	3	—	372	61	3	—	3	412	13	7	—	—	452	3	—	3	7
333	—	3	47	3	373	7	—	37	—	413	3	—	3	—	453	23	3	13	3
334	13	—	—	17	374	3	19	3	23	414	41	3	11	3	454	19	7	—	—
335	3	7	3	—	375	11	3	13	3	415	7	—	—	—	455	3	29	3	47
336	—	3	7	3	376	—	53	—	—	416	3	23	3	11	456	—	3	—	3
337	—	—	11	31	377	3	7	3	—	417	43	3	—	3	457	7	17	23	19
338	3	17	3	—	378	19	3	7	3	418	37	47	53	59	458	3	—	3	13
339	—	3	43	3	379	17	—	—	29	419	3	7	3	13	459	—	3	—	3
340	19	41	—	7	**380**	3	—	3	31	**420**	—	3	7	3	**460**	43	—	17	11
341	3	—	3	13	381	37	3	11	3	421	—	11	—	—	461	3	7	3	31
342	11	3	23	3	382	—	—	43	7	422	3	41	3	—	462	—	3	7	3
343	47	—	7	19	383	3	—	3	11	423	—	3	19	3	463	11	41	—	—
344	3	11	3	—	384	23	3	—	3	424	—	—	31	7	464	3	—	3	—
345	7	3	—	3	385	—	—	7	17	425	3	—	3	—	465	—	3	—	3
346	—	—	—	—	386	3	—	3	53	426	—	3	17	3	466	59	—	13	7
347	3	23	3	7	387	7	3	—	3	427	—	—	7	11	467	3	—	3	—
348	59	3	11	3	388	—	11	13	—	428	3	—	3	—	468	31	3	43	3
349	—	7	13	—	389	3	17	3	7	429	7	3	—	3	469	—	13	7	37
350	3	31	3	11	**390**	47	3	—	3	**430**	11	13	59	31	**470**	3	—	3	17
351	—	3	—	3	391	—	7	—	—	431	3	19	3	7	471	7	3	53	3
352	7	13	—	—	392	3	—	3	—	432	29	3	—	3	472	—	—	29	—
353	3	—	3	—	393	—	3	31	3	433	61	7	—	—	473	3	—	3	7
354	—	3	—	3	394	7	—	—	11	434	3	43	3	—	474	11	3	47	3
355	53	11	—	—	395	3	59	3	37	435	19	3	—	3	475	—	7	67	—
356	3	7	3	43	396	17	3	—	3	436	7	—	11	17	476	3	11	3	19
357	—	3	7	3	397	11	29	41	23	437	3	—	3	29	477	13	3	17	3
358	—	—	17	37	398	3	7	3	—	438	13	3	41	3	478	7	—	—	—
359	3	—	3	59	399	13	3	7	3	439	—	23	—	53	479	3	—	3	—

	1	3	7	9		1	3	7	9		1	3	7	9		1	3	7	9
480	—	3	11	3	**520**	7	11	41	—	**560**	3	13	3	71	**600**	17	3	—	3
481	17	—	—	61	521	3	13	3	17	561	31	3	41	3	601	—	7	11	13
482	3	7	3	11	522	23	3	—	3	562	7	—	17	13	602	3	19	3	—
483	—	3	7	3	523	—	—	—	13	563	3	43	3	—	603	37	3	—	3
484	47	29	37	13	524	3	7	3	29	564	—	3	—	3	604	7	—	—	23
485	3	23	3	43	525	59	3	7	3	565	—	—	—	—	605	3	—	3	73
486	—	3	31	3	526	—	19	23	11	566	3	7	3	—	606	11	3	—	3
487	—	11	—	7	527	3	—	3	—	567	53	3	7	3	607	13	—	59	—
488	3	19	3	—	528	—	3	17	3	568	13	—	11	—	608	3	7	3	—
489	67	3	59	3	529	11	67	—	7	569	3	—	3	41	609	—	3	7	3
490	13	—	7	—	**530**	3	—	3	—	**570**	—	3	13	3	**610**	—	17	31	41
491	3	17	3	—	531	47	3	13	3	571	—	29	—	7	611	3	—	3	29
492	7	3	13	3	532	17	—	7	73	572	3	59	3	17	612	—	3	11	3
493	—	—	—	11	533	3	—	3	19	573	11	3	—	3	613	—	—	17	7
494	3	—	3	7	534	7	3	—	3	574	—	—	7	—	614	3	—	3	11
495	—	3	—	3	535	—	53	11	23	575	3	11	3	13	615	—	3	47	3
496	11	7	—	—	536	3	31	3	7	576	7	3	73	3	616	61	—	7	31
497	3	—	3	13	537	41	3	19	3	577	29	23	53	—	617	3	—	3	37
498	17	3	—	3	538	—	7	—	17	578	3	—	3	7	618	7	3	23	3
499	7	—	19	—	539	3	—	3	—	579	—	3	11	3	619	41	11	—	—
500	3	—	3	—	**540**	11	3	—	3	**580**	—	7	—	37	**620**	3	—	3	7
501	—	3	29	3	541	7	—	—	—	581	3	—	3	11	621	—	3	—	3
502	—	—	11	47	542	3	11	3	61	582	—	3	—	3	622	—	7	13	—
503	3	7	3	—	543	—	3	—	3	583	7	19	13	—	623	3	23	3	17
504	71	3	7	3	544	—	—	13	—	584	3	—	3	—	624	79	3	—	3
505	—	31	13	—	545	3	7	3	53	585	—	3	—	3	625	7	13	—	11
506	3	61	3	37	546	43	3	7	3	586	—	11	—	—	626	3	—	3	—
507	11	3	—	3	547	—	13	—	—	587	3	7	3	—	627	—	3	—	3
508	—	13	—	7	548	3	—	3	11	588	—	3	7	3	628	11	61	—	19
509	3	11	3	—	549	17	3	23	3	589	43	71	—	17	629	3	7	3	—
510	—	3	—	3	**550**	—	—	—	7	**590**	3	—	3	19	**630**	—	3	7	3
511	19	—	7	—	551	3	37	3	—	591	23	3	61	3	631	—	59	—	71
512	3	47	3	23	552	—	3	—	3	592	31	—	—	7	632	3	—	3	—
513	7	3	11	3	553	—	11	7	29	593	3	17	3	—	633	13	3	—	3
514	53	37	—	19	554	3	23	3	31	594	13	3	19	3	634	17	—	11	7
515	3	—	3	7	555	7	3	—	3	595	11	—	7	59	635	3	—	3	—
516	13	3	—	3	556	67	—	19	—	596	3	67	3	47	636	—	3	—	3
517	—	7	31	—	557	3	—	3	7	597	7	3	43	3	637	23	—	7	—
518	3	71	3	—	558	—	3	37	3	598	—	31	—	53	638	3	13	3	—
519	29	3	—	3	559	—	7	29	11	599	3	13	3	7	639	7	3	—	3

	1 3 7 9		1 3 7 9		1 3 7 9		1 3 7 9
640	37 19 43 13	**680**	3 — 3 11	**720**	19 3 — 3	**760**	11 — — 7
641	3 11 3 7	681	7 3 17 3	721	— — 7 —	761	3 23 3 19
642	— 3 — 3	682	19 — — —	722	3 31 3 —	762	— 3 29 3
643	59 7 41 47	683	3 — 3 7	723	7 3 — 3	763	13 17 7 —
644	3 17 3 —	684	— 3 41 3	724	13 — — 11	764	3 — 3 —
645	— 3 11 3	685	13 7 — 19	725	3 — 3 7	765	7 3 13 3
646	7 23 29 —	686	3 — 3 —	726	53 3 13 3	766	47 79 11 —
647	3 — 3 11	687	— 3 13 3	727	11 7 19 29	767	3 — 3 7
648	— 3 13 3	688	7 — 71 83	728	3 — 3 37	768	— 3 — 3
649	— 43 73 67	689	3 61 3 —	729	23 3 — 3	769	— 7 43 —
650	3 7 3 23	**690**	67 3 — 3	**730**	7 67 — —	**770**	3 — 3 13
651	17 3 7 3	691	— 31 — 11	731	3 71 3 13	771	11 3 — 3
652	— 11 61 —	692	3 7 3 13	732	— 3 17 3	772	7 — — 59
653	3 47 3 13	693	29 3 7 3	733	— — 11 41	773	3 11 3 71
654	31 3 — 3	694	11 53 — —	734	3 7 3 —	774	— 3 61 3
655	— — 79 7	695	3 17 3 —	735	— 3 7 3	775	23 — — —
656	3 — 3 —	696	— 3 — 3	736	17 37 53 —	776	3 7 3 17
657	— 3 — 3	697	— 19 — 7	737	3 73 3 47	777	19 3 7 3
658	— 29 7 11	698	3 — 3 29	738	11 3 83 3	778	31 43 13 —
659	3 19 3 —	699	— 3 — 3	739	19 — 13 7	779	3 — 3 11
660	7 3 — 3	**700**	— 47 7 43	**740**	3 11 3 31	**780**	29 3 37 3
661	11 17 13 —	701	3 — 3 —	741	— 3 — 3	781	73 13 — 7
662	3 37 3 7	702	7 3 — 3	742	41 13 7 17	782	3 — 3 —
663	19 3 — 3	703	79 13 31 —	743	3 — 3 43	783	41 3 17 3
664	29 7 17 61	704	3 — 3 7	744	7 3 11 3	784	— 11 7 47
665	3 — 3 —	705	11 3 — 3	745	— 29 — —	785	3 — 3 29
666	— 3 59 3	706	23 7 37 —	746	3 17 3 7	786	7 3 — 3
667	7 — 11 —	707	3 11 3 —	747	31 3 — 3	787	17 — — —
668	3 41 3 —	708	73 3 19 3	748	— 7 — —	788	3 — 3 7
669	— 3 37 3	709	7 41 47 31	749	3 59 3 —	789	13 3 53 3
670	— — 19 —	**710**	3 — 3 —	**750**	13 3 — 3	**790**	— 7 — 11
671	3 7 3 —	711	13 3 11 3	751	7 11 — 73	791	3 41 3 —
672	11 3 7 3	712	— 17 — —	752	3 — 3 —	792	89 3 — 3
673	53 — — 23	713	3 7 3 11	753	17 3 — 3	793	7 — — 17
674	3 11 3 17	714	37 3 7 3	754	— 19 — —	794	3 13 3 —
675	43 3 29 3	715	— 23 17 —	755	3 7 3 —	795	— 3 73 3
676	— — 67 7	716	3 13 3 67	756	— 3 7 3	796	19 — 31 13
677	3 13 3 —	717	71 3 — 3	757	67 — — 11	797	3 7 3 79
678	— 3 11 3	718	43 11 — 7	758	3 — 3 —	798	23 3 7 3
679	— — 7 13	719	3 — 3 23	759	— 3 71 3	799	61 — 11 19

	1	3	7	9		1	3	7	9		1	3	7	9		1	3	7	9
800	3	53	3	—	**840**	31	3	7	3	**880**	13	—	—	23	**920**	3	—	3	—
801	—	3	—	3	841	13	47	19	—	881	3	7	3	—	921	61	3	13	3
802	13	71	23	7	842	3	—	3	—	882	—	3	7	3	922	—	23	—	11
803	3	29	3	—	843	—	3	11	3	883	—	11	—	—	923	3	7	3	—
804	11	3	13	3	844	23	—	—	7	884	3	37	3	—	924	—	3	7	3
805	83	—	7	—	845	3	79	3	11	885	53	3	17	3	925	11	19	—	47
806	3	11	3	—	846	—	3	—	3	886	—	—	—	7	926	3	59	3	13
807	7	3	41	3	847	43	37	7	61	887	3	19	3	13	927	73	3	—	3
808	—	59	—	—	848	3	17	3	13	888	83	3	—	3	928	—	—	37	7
809	3	—	3	7	849	7	3	29	3	889	17	—	7	11	929	3	—	3	17
810	—	3	11	3	**850**	—	11	47	67	**890**	3	29	3	59	**930**	71	3	41	3
811	—	7	—	23	851	3	—	3	7	891	7	3	37	3	931	—	67	7	—
812	3	—	3	11	852	—	3	—	3	892	11	—	79	—	932	3	—	3	19
813	47	3	79	3	853	19	7	—	—	893	3	—	3	7	933	7	3	—	3
814	7	17	—	29	854	3	—	3	83	894	—	3	23	3	934	—	—	13	—
815	3	31	3	41	855	17	3	43	3	895	—	7	13	17	935	3	47	3	7
816	—	3	—	3	856	7	—	13	11	896	3	—	3	—	936	11	3	17	3
817	—	11	13	—	857	3	—	3	23	897	—	3	47	3	937	—	7	—	83
818	3	7	3	19	858	—	3	31	3	898	7	13	11	89	938	3	11	3	41
819	—	3	7	3	859	11	13	—	—	899	3	17	3	—	939	—	3	—	3
820	59	13	29	—	**860**	3	7	3	—	**900**	—	3	—	3	**940**	7	—	23	97
821	3	43	3	—	861	79	3	7	3	901	—	—	71	29	941	3	—	3	—
822	—	3	19	3	862	37	—	—	—	902	3	7	3	—	942	—	3	11	3
823	—	—	—	7	863	3	89	3	53	903	11	3	7	3	943	—	—	—	—
824	3	—	3	73	864	—	3	—	3	904	—	—	83	—	944	3	7	3	11
825	37	3	23	3	865	41	17	11	7	905	3	11	3	—	945	13	3	7	3
826	11	—	7	—	866	3	—	3	—	906	13	3	—	3	946	—	—	—	17
827	3	—	3	17	867	13	3	—	3	907	47	43	29	7	947	3	—	3	—
828	7	3	—	3	868	—	19	7	—	908	3	31	3	61	948	19	3	53	3
829	—	—	—	43	869	3	—	3	—	909	—	3	11	3	949	—	11	—	7
830	3	19	3	7	**870**	7	3	—	3	**910**	19	—	7	—	**950**	3	13	3	37
831	—	3	—	3	871	31	—	23	—	911	3	31	3	11	951	—	3	31	3
832	53	7	11	—	872	3	11	3	7	912	7	3	—	3	952	—	89	7	13
833	3	13	3	31	873	—	3	—	3	913	23	—	—	13	953	3	—	3	—
834	19	3	17	3	874	—	7	—	13	914	3	41	3	7	954	7	3	—	3
835	7	—	61	13	875	3	—	3	19	915	—	3	—	3	955	—	41	19	11
836	3	—	3	—	876	—	3	11	3	916	—	7	89	53	956	3	73	3	7
837	11	3	—	3	877	7	31	67	—	917	3	—	3	67	957	17	3	61	3
838	17	83	—	—	878	3	—	3	11	918	—	3	—	3	958	11	7	—	43
839	3	7	3	37	879	59	3	19	3	919	7	29	17	—	959	3	53	3	29

	1	3	7	9		1	3	7	9		1	3	7	9		1	3	7	9
960	—	3	13	3	**970**	89	31	18	7	**980**	3	—	3	17	**990**	—	3	—	3
961	7	—	59	—	971	3	11	3	—	981	—	3	—	3	991	11	23	47	7
962	3	—	3	—	972	—	3	71	3	982	7	11	31	—	992	3	—	3	—
963	—	3	23	3	973	37	—	7	—	983	3	—	3	—	993	—	3	19	3
964	31	—	11	—	974	3	—	3	—	984	13	3	43	3	994	—	61	7	—
965	3	7	3	13	975	7	3	11	3	985	—	59	—	—	995	3	37	3	23
966	—	3	7	3	976	43	13	—	—	986	3	7	3	71	996	7	3	—	3
967	19	17	—	—	977	3	29	3	7	987	—	3	7	3	997	13	—	11	17
968	3	23	3	—	978	—	3	—	3	988	41	—	—	11	998	3	67	3	7
969	11	3	—	3	979	—	7	97	41	989	3	13	3	19	999	97	3	13	3

References

This text gives an elementary introduction to elementary number theory. It does not go as far, as fast, or as deep as many other books. If you wish to go further, there are many places you can go. The best book for further browsing in number theory is the classic

G. H. Hardy and E. M. Wright, *The Theory of Numbers,* 4th ed. (New York: Oxford University Press, 1960).

There is no better place to get the flavor of number theory than in this elegantly written book. Though not a text, it gives introductions to many branches of number theory. The subjects it covers range widely in level of difficulty, from how to win at Nim to a proof of the Prime Number Theorem.

There are a number of superior number theory texts at a higher level than this one. Seven recent ones are

G. E. Andrews, *Number Theory* (Philadelphia: Saunders, 1971).

R. G. Archibald, *An Introduction to the Theory of Numbers* (Columbus, Ohio: Merrill, 1970).

H. Davenport, *The Higher Arithmetic* (reprinted, New York: Harper, 1970).

Emil Grosswald, *Topics from the Theory of Numbers,* (New York: Macmillan, 1966).

Ivan Niven and H. S. Zukerman, *An Introduction to the Theory of Numbers,* 3rd ed. (New York: Wiley, 1972).

Joe Roberts, *Elementary Number Theory, A Problem Oriented Approach* (Cambridge, Mass.: M.I.T., 1977).

H. M. Stark, *An Introduction to Number Theory* (Chicago: Markham, 1970).

Four books that are unusual for their richness are

Donald Knuth, *The Art of Computer Programming* (vol. 1, *Fundamental Algorithms,* and vol. 2, *Seminumerical Algorithms* (Reading, Mass.: Addison-Wesley, 1968, 1969).

Waclaw Sierpinski, *Elementary Theory of Numbers* (Warsaw: Panstowowe Wydawnictwo Naukowe, 1964).

Daniel Shanks, *Solved and Unsolved Problems in Number Theory* (New York: Spartan, 1962).

Many texts contain material on continued fractions, which is not really necessary since the job is done so well by

C. D. Olds, *Continued Fractions* (New York: Random House, 1963).

Noteworthy among popular works is

A. H. Beiler, *Recreations in the Theory of Numbers* (New York: Dover, 1964).

Beiler is crazy about numbers and he reports on various results and calculations with great enthusiasm.

This list could be extended to almost any given length, one lifetime is not time enough to absorb all of number theory. All that can be done is to make a start on it; you have done that, and can do more.

There follows a list of the works referred to in the text.

1. I. A. Barnett, "The theory of numbers as a required course in the college curriculum for majors." *American Mathematical Monthly* **73**(1966): 1002–1004.

2. Albert Beiler, *Recreations in the Theory of Numbers* (New York: Dover, 1964).

3. E. T. Bell, *The Last Problem* (New York: Simon and Schuster, 1961).

4. Martin Davis, "Hilbert's tenth problem is unsolvable." *American Mathematical Monthly,* **80**(1973):233–269.

5. J. Fang, *Hilbert* (Hauppauge, N.Y.: Paiedeia, 1970).

6. J. P. Jones, D. Sato, H. Wada, and D. Wiens, "Diophantine representation of the set of prime numbers." *American Mathematical Monthly,* **83**(1976):449–464.

7. Edgar Karst, "12 to 16 primes in arithmetic progression." *Journal of Recreational Mathematics,* **2**(1969):214–215.

8. Elisha S. Loomis, *The Pythagorean Proposition.* (Reprinted, Reston, Va.: National Council of Teachers of Mathematics, 1968).

9. *Mathematical Reviews,* **41**(1971):#5282.

10. *Mathematical Reviews,* **45**(1973):#1830.

11. W. H. Mills, "A prime-representing function," *Bulletin of the American Mathematical Society,* **53**(1947):604.

12. Leo Moser, "On the theorems of Wilson and Fermat." *Scripta Mathematica,* **22**(1956):288.

13. C. S. Ogilvy, and J. Anderson, *Excursions in Number Theory* (New York: Oxford, 1966).

14. Joe Roberts, *Elementary Number Theory* (Cambridge, Mass.: M.I.T., 1977).

15. William L. Schaaf, *A Bibliography of Recreational Mathematics,* 3 vols. (Reston, Va.: National Council of Teachers of Mathematics, 1955, 1970, 1973).

Answers to Selected Exercises

Section 1

1. All of them.

4. 2, 5, 2.

5. 1, n.

6. d.

7. 3, 3; 3, 0.

9. 7, 34.

Section 2

1. One, one.

3. $72 = 2^3 \cdot 3^2$, $480 = 2^5 \cdot 3 \cdot 5$.

4. 25, 45, 65, 81, and 85.

5. $2 \cdot 3 \cdot 5^2 \cdot 53$.

Section 3

1. The left-hand side is even; the right-hand side is odd.

2. All solutions are $x = 5t$, $y = 2 - t$, t an integer.

3. (c).

4. $x = 10 + 3t$, $y = -t$, t an integer.

5. $x = 6$, $y = 1$ and $x = 3$, $y = 2$.

Section 4

1. True, true, false, true.

2. Then $km = a - b$, so $m \,|\, (a - b)$ and $a \equiv b \pmod{m}$.

3. 1, 7, 9, 4, and 8.

4. $n \equiv 1 \pmod 2$. $n = 1 + 2k$ for some k. n leaves a remainder of 1 when divided by 2.

10. For example, $5 \cdot 4 \equiv 5 \cdot 6 \pmod{10}$, but $4 \not\equiv 6 \pmod{10}$.

11. (a) $x \equiv 2 \pmod 7$, (b) $x \equiv 4 \pmod 7$.

12. $x \equiv 2 \pmod 3$.

Section 5

1. For example, $4x \equiv 3$, $5x \equiv 4$, $6x \equiv 6 \pmod{12}$ have, respectively, 0, 1, and 6 solutions.

2. (b), (c), and (d) have no solutions.

3. This is included in Theorem 1.

4. (a) 2, (b) $x = -1 + 10t$, $y = 2 - 9t$.

5. 3, 1, 5, 0, 1.

6. $x = 2, 5, 8, 11,$ or 14.

7. 2, 7, 12; 2; none; 2.

Section 6

2. 10, 1. **3.** (2, 6), (3, 4), (5, 9), (7, 8).

Section 7

1.

n	11	12	13	14	15	16
$d(n)$	2	6	2	4	4	5.

2. $d(p^3) = 4.\ d(p^n) = n + 1.$

3. $d(p^3q) = 8.\ d(p^nq) = 2(n + 1).$

4. 20.

5.

n	9	10	11	12	13	14
$\sigma(n)$	13	18	12	28	14	24

6. $\sigma(p^3) = 1 + p + p^2 + p^3 \cdot \sigma(pq) = 1 + p + q + pq.$

8. $\sigma(p^n) = 1 + p + p^2 + \cdots + p^n = (p^{n+1} - 1)/(p - 1).$

9. $\sigma(240) = 744.$

10.

n	13	14	15	16	17	18	19	20	21	22	23	24
$f(n)$	1	1	1	32	1	6	1	4	1	1	1	12

Section 9

5. 1, 3. 1, 3, 5, 7. 1, 3, 5, 7, 9, 11, 13, 15. $\phi(2^n) = 2^{n-1}$, the number of odd positive integers less than 2^n.

7. $\phi(m)$.

8. 24, 36, 36.

9. (a) 12, 13, 14, 15, 16. (b) 2^k. (c) p^k.

10. $C_1 = \{1, 3, 5, 9, 11, 13\}$, $C_2 = \{2, 4, 6, 8, 10, 12\}$, $C_7 = \{7\}$, $C_{14} = \{14\}$.

Section 10

1. 2, 2, and 2.
2. 1, 2, 3, or 6. 2 and 5 have order six, 4 and 7 have order three, 8 has order two, and 1 has order one.
3. 191 (39, 77, 115, and 153 are composite).
5. 3 and 7 are primitive roots of 10.

Section 11

1. $x^2 + 4x + 3 \equiv 0$ (mod 5).
2. $(x + 2)^2 \equiv 1$ (mod 5).
3. 2 and 4.
4. 2 and $p - 2$.
5. 1, 2, and 4.
6. 15 and 16.

7. 1, 1, 1, 1.
8. 1, 1, 1.
9. If $p \nmid a$, then $(a^2/p) = 1$.
13. 1, 1.
14. 5, 13, 17.
15. $-1, -1$.

Section 12

2. No solution.

Section 13

1. $31 = 2^4 + 2^3 + 2^2 + 2^1 + 2^0$. $33 = 2^5 + 2^0$.
2. 6, 7.
3. $n' < 2^r$. Hence $r > e_i$ for all i.
4. 9, 7, 64.
5. 10_2, 10100_2, 11001000_2.

Section 14

1. 11, 19, 110, 6χ.
3. 28, 40, $6\chi6$.
4. 8.
5. 371, 275.
6. $.\overline{186\chi35}$

Section 15

1. 73/4950.
2. $.\overline{17073}$.
3. 4, 2, 2.

4. 7, 6, 5.
5. $.\overline{02439}$.
6. 5.

Section 16

1. If p divides any two of x, y, and z, then it divides the third.

2. Because 2 would divide (a, b).

Section 17

1. $c^2 \equiv 2 \pmod 4$ is impossible.

6. Because $n = 2v^2$. Because $b^2 = m^2 - n^2$.

7. If $n = 0$, then $a^2 = 2mm = 0$, and a, b, c would be a trivial solution.

Section 18

2. $325 = 18^2 + 1^2$.

Section 19

1. $3 \cdot 17 = 7^2 + 1^2 + 1^2 + 0^2 = 5^2 + 4^2 + 3^2 + 1^2$.

Section 20

1. The smallest solutions are $3^2 - 2 \cdot 2^2 = 1$ and $2^2 - 3 \cdot 1^2 = 1$.

2. If $x - my = x + my = 1$, then $2x = 2$. The other case is similar.

3. 99, 35.

4. $(r + sN^{1/2})^k (r - sN^{1/2})^k = (r^2 - Ns^2)^k = 1^k = 1$.

5. Let $a = c = x_1$ and $b = d = y_1$ in Lemma 1.

6. 7, 4 and 26, 15.

Section 21

1. 1, 9, 8, 4.

3. 10, 5.

2. 20, 70.

4. 1979.

Appendix A

1. $f(n) = (n - 1)^2$.

4. All positive even integers.

2. $f(n) = n^2 + 1$.

5. "$1 = 1 \cdot 2/2$." Yes.

3. $f(n) = n^2 - 6n + 7$.

Answers to Selected Odd-Numbered Problems

Section 1

1. 1 and 592.

3. One solution is $x = -40$, $y = 79$.

11. (b) Yes.

Section 2

1. $2 \cdot 617$, $2^8 \cdot 3^3 \cdot 5$, $3 \cdot 7 \cdot 11 \cdot 13 \cdot 37$.

7. 60466176.

9. No.

11. (b) No.

13. False.

15. No.

Section 3

1. $x = 1 + t$, $y = 1 - t$; $x = 3 + 4t$, $y = 1 + 3t$; $x = -1 + 16t$, $y = 2 - 15t$; t an integer.

3. $x = 1$, $y = 1$; $x = 3 + 4t$, $y = 1 + 3t$, $t = 0, 1, \ldots$; $x = 1$, $y = 6$ and $x = 3$, $y = 1$.

5. $(x, y, z) = (22, 8, 1)$, $(23, 6, 2)$, $(24, 4, 3)$, or $(25, 2, 4)$.

7. Nine apples at 9¢ each and three oranges at 6¢ each.

9. If the first merchant had d, the second had $3d$, the third $5d$, and the purse $15d$.

Section 4

1. 0, 2, 78.
5. 1, 2, 5, 10, 71, 142, 355, or 710.
7. 3.

11. 6.
13. 0, 1, 5, or 6
17. A is.

Section 5

1. 9; 6; 2, 8, 14; 1752.
3. $x \equiv 1 \pmod 6$, $x \equiv 348 \pmod{385}$, $x \equiv 3 \pmod{385}$.
5. 0, 1, 2, 4, 5, 10, or 20 solutions.
7. 1292.
9. $x = 3, y = 0; x = 5, y = 5$.
11. 17, 157, or other larger and more unlikely numbers.
13. 213.
15. 30233088000000.
19. $(m_i, m_j) | (a_i - a_j)$ for all i and j, $i \neq j$.

Section 6

1. 1, 4, 1.
3. 3.
5. 1.

7. 101
19. All n such that $n \not\equiv 0$ or $1 \pmod p$.

Section 7

1. 8, 96, 24, 1344.
3. 12, 14736; 4, 120144.
7. 24, 48.

9. There are three: 5040, 7920, and 9360.
11. Even k.
17. $2d(N)$.

Section 8

3. 12, 18, and 20 are abundant, 6 is perfect, and the rest are deficient.

Section 9

1. 12, 96, 960.
3. 6528, 80088.

5.

a	1	2	3	4	5	6	7	8	9	10	11	12	13	14
a^8 (mod 15)	1	1	6	1	10	6	1	1	6	10	1	6	1	1

15. 5, 8, and 12.

Section 10

1. 2, 6, 7, and 11 have order 12; 4 and 10 have order 6; 3 and 9 have order 3; 12 has order 2; 1 has order 1.

3. They are 3, 10, 13, 14, and 15.

5. 4, 2, 4, 4, 2, 4, and 2. No.

7. 6 and 26.

15. $p - 4$.

Section 11

1. All of them.

3. $x \equiv 22$ or 31 (mod 53), $x \equiv 13$ or 18 (mod 31), $x \equiv 2$ or 5 (mod 7), $x \equiv 5$ or 992 (mod 997).

5. $-1, -1, -1, 1$.

7. $x \equiv 3$ or 6 (mod 7), $x \equiv 50$ or 100 (mod 101).

9. 1, 1.

17. Yes: 23, 76, 83, and 136 are all solutions.

19. $(r/p) = 1$.

Section 12

1. $(3/p) = 1$ if $p \equiv 1$ or 11 (mod 12) and -1 if $p \equiv 5$ or 7 (mod 12).

7. (b) Then a and $p - a$ are both residues or nonresidues.

9. -1.

Section 13

1. 10111010100_2, 2001021_3, 4226_7, 2037_9, 1137_{11}.

3. 421, 1107, 4159, 5377.

5. 102, 153, 7.

7.

\cdot	2	3	4	5	6
2	4	6	11	13	15
3	6	12	15	21	24
4	11	15	22	26	33
5	13	21	26	34	42
6	15	24	33	42	51

9. 73, 156, 214, 48848.

11. 19/49, 1/2, 13/16.

13. Any $b \equiv 2, 3,$ or 4 (mod 5).

15. Six gifts, one each of \$32, \$128, \$512, \$1024, \$32768, and \$65536.

17. (a) COl, DODO, CODA, BODE.
 (b) 3243, 45232, 57007, 41438.
 (c) Not that I know of.

19. (a) 296, 1968.
 (b) (113110)₁.

Section 14

1. 8χ67, 15є43126.

3. .572497249 . . . , 3/13.

5. $4є.є6.

є. The digits in this answers are in the base χ. One do-metric mile is .98 . . . ordinary miles, one do-metric pint is .93 . . . ordinary pints, and one do-metric pound is .98 . . . ordinary pounds.

Section 15

1. 2, 1, 4.

3. 2, 3.

7. 1/16, 1/18, and 1/24.

9. 2, 4, 3, 6, 10.

11. 6, 1, 16.

13. (a) $\overline{.01234579}$ (b) $\overline{.0123457}$
 (c) .012346.

Section 16

1. They have hypotenuses 10, 15, 20, 25, 26, 30, 34, 35, 39, 40, and 45.

3. (100, 2499, 2501) and (100, 621, 625).

9. Yes.

13. Only (40, 9, 41).

Section 17

3. No.

5. No.

7. No.

9. All even k.

Section 18

1. None is a sum of two squares.

3. The larger square is 179, 178, 173, 166, 163, 157, 142, or 131.

7. Yes.

Section 19

1. $31 = 5^2 + 2^2 + 1^2 + 1^2, 37 = 6^2 + 1^2 = 4^2 + 4^2 + 2^2 + 1^2, 41 = 6^2 + 2^2 + 1^2 = 5^2 + 4^2 = 4^2 + 4^2 + 3^2, 43 = 5^2 + 4^2 + 1^2 + 1^2 = 5^5 + 3^2 + 3^2, 47 = 6^2 + 3^2 + 1^2 + 1^2 = 5^2 + 3^2 + 3^2 + 2^2, 53 = 7^2 + 2^2 = 6^2 + 4^2 + 1^2$.

3. $148^2 + 108^2 + 12^2 + 4^2$.

5. $0^2 + 7^2 \equiv 8^2 + 6^2 \equiv 4^2 + 4^2 \equiv -2 \pmod{17}$.

Section 20

1. $3 + 8^{1/2}, 5 + 2 \cdot 6^{1/2}, 7 + 2 \cdot 12^{1/2}, 19 + 6 \cdot 10^{1/2}$.

3. $(3, 1)$ and $(17, 6)$, $(5, 2)$ and $(49, 20)$, $(8, 1)$ and $(127, 16)$.

5. The three smallest positive solutions are $(1, 1)$, $(3, 4)$, and $(11, 15)$.

7. (c) There are infinitely many solutions.

9. (b) Triangles with sides $(3, 4, 5)$, $(13, 14, 15)$, and $(51, 52, 53)$ are the smallest.

13. x_k/y_k approaches $n^{1/2}$.

15. The next solution is $108^2 + 109^2 + 110^2 = 133^2 + 134^2$.

Section 21

1. (a) 1979. (b) None.

Section 23: Additional Problems

Section 1

1. All solutions are $x = 1 + 19t$, $y = 1 - 23t$, t an integer.

7. $(n, n + 20) = 1, 2, 4, 5, 10,$ or 20.

9. Yes.

Section 2

1. In its prime-power decomposition each exponent is a multiple of k.

3. (a) $60, 2p^2q$.

Section 3

1. Four ways. There may be 14, 15, 16, or 17 quarters.

Section 4

3. (a) 0, 1, 4, or 7. (b) No.

7. Any $n \equiv 15 \pmod{30}$ satisfies the three congruences.

Section 5

1. $x \equiv 534 \pmod{2401}$. 3. $k = 1$ or 2.

Section 7

3. (c) 18, 20, and 24 are examples.

5. (b) No.

Section 8

7. All $a \geq 3$.

Section 9

3. All n which are a power of 2 times an integer relatively prime to 3.

7. (a) 1, 1, 2, 1, 2, 2, 2. (b) $k - 1$. (c) k. (d) $j + k - 1$.

13. (a) 0, -13, 0, -15, 0. (b) $-p$. (c) 0. (d) $-p^k$.

15. All such $2k, k < 50$, are 14, 26, 34, 38, 50, 62, 68, 74, 76, 86, 90, 94, and 98.

Section 10

1. None.

3. 17.

5. $1983^{512} \equiv 1$ (mod 1024) and 1983 is not a multiple of 512.

9.

k	1	2	3	4	5	6	7	8	9	10
$\mathrm{ind}_2 k$ (mod 19)	0	1	13	2	16	14	6	3	8	17

k	11	12	13	14	15	16	17	18
$\mathrm{ind}_2 k$ (mod 19)	12	15	5	7	11	4	10	9

Section 11

1. No solutions, $x \equiv 0$ or 4 (mod 5), $x \equiv 2$ (mod 5).

3. 1, 2, 4, 7, 8, 9, 10, 14, 16, 18, 19, 20, 25, 28.

5. 3, 11, and 17.

7. Yes: 21 or 76. Yes: 16 or 37.

9. No.

Section 12

1. No. There are no longer sequences.

3. (c) 2, 3, and 8.

Section 13

1. (a) $(a, b) = (2, 16)$ or $(1, 33)$. (b) $(a, b) = (6, 17)$.

Section 15

1. $\phi(31415) \neq 31414$.

3. In any base b such that 2, 3, and 5 divide b.

5. A computer program may be necessary to find the next example, even though it comes before 1/100: 1/77 and 1/78 both have period 6.

Section 17

7. Yes.

9. $x = (ac)^{rn}, y = (bc)^{rn}, z = c^s$, where $c = a^{rn^2} + b^{rn^2}$ and $rn^2 + 1 = (n - 1)s$.

Section 18

9. None.

Section 19

1. $2387^2 + 154^2 + 66^2 + 0^2$, among others.

Section 20

1. Approximately .09, .002, .00007, and .000002.

5. 3363/2378.

Miscellaneous

5. Eight 13's, two 9's, and seventy-eight 3's.

7. Two scotch, one rum, and four vodka.

17. 32976.

19. 5.

23. $533 = 13 \cdot 41$, $1073 = 29 \cdot 37$.

27. No.

31. (a) $a = 7, p = 17$.

41. $(c, d) = (4, 3)$ or $(11, 9)$.

49. 1996, 2024, 2052, and 2080.

51. No.

53. $2d(N^2) = 1$.

55. (a) $p - 2$. (b) $p^2 - 2p$.

65. $x = -1 - 9m + 8n, y = 1 + 9m - 7n, z = -m$ is one way of writing the solutions.

73. Yes.

81. For example, $2^2 + 5^2 + \cdots + 23^2 + 26^2 = 48^2$.

83. 1, 2, 3, 5, 7, and 11.

85. $9 = (12/5)^2 + (9/5)^2 = (36/13)^2 + (15/13)^2 = (24/17)^2 + (45/17)^2 = \ldots$.

97. (b) Yes: 1806. (c) No others.

Appendix A

13. No.

Comments on Selected Odd-Numbered Problems

The outlines of solutions that follow may not be the easiest or the quickest, but they will almost always work. The hints may not convey anything, but they are almost never misleading.

Section 2

7. $60466176 = 6^{10} = (6^5)^2 = (6^2)^5$.

11. (b) $78 \le (25ab)/32 \le 82$ no matter what the digits a and b are, and none of 78, 79, 80, 81, and 82 is a power, except 81.

13. A counterexample is: $3 | 60$, $5 | 60$, and both 3 and 5 are greater than $60^{1/4} = 2.78 \ldots$. But $60/3 \cdot 5 = 4$ is not prime. The statement would be true if p and q were the two smallest prime factors of n.

15. $2^{11} - 1 = 2047 = 23 \cdot 89$ is composite and 11 is not.

Section 5

1. $40x \equiv 777 \equiv -1000 \pmod{1777}$, so $x \equiv -25 \equiv 1752 \pmod{1777}$.

15. $2^{15} \cdot 3^{10} \cdot 5^6$.

19. The condition $(m_i, m_j) | (a_i - a_j)$ for all i and j, $i \ne j$, is both necessary and sufficient for the system to have a solution.

Section 6

11. $-1 \equiv (p - 1)! \equiv (p - 1)(p - 2)(p - 3)! \equiv (-1)(-2)(p - 3)! \equiv 2(p - 3)!$ \pmod{p}.

15. (b) $1^p + 2^p + \cdots + (p - 1)^p \equiv 1 + 2 + \cdots + (p - 1) \equiv p(p - 1)/2 \equiv 0$ \pmod{p}, since $(p - 1)/2$ is an integer.

17. (a) $a^{p+q} - a^{p+1} - a^{q+1} + a^2 = (a^p - a)(a^q - a)$.

(b) $a^{pq} - a^p \equiv a^q - a \pmod{p}$ and $a^{pq} - a^q \equiv a^p - a \pmod{q}$, so both p and q divide $a^{pq} - a^p - a^q + a$.

19. If $n \equiv 0 \pmod{p}$, then $1 + n + n^2 + \cdots + n^{p-2} \equiv 1 \pmod{p}$. If $n \equiv 1 \pmod{p}$, the sum is congruent to $p - 1 \pmod{p}$. If $n \not\equiv 0$ or $1 \pmod{p}$, the sum is $(n^{p-1} - 1)/(n - 1)$, and p divides the numerator but not the denominator of the fraction. Thus p divides the sum.

Section 7

15. If d_1, d_2, \ldots, d_k are the divisors of n, then $1/d_1 + 1/d_2 + \cdots + 1/d_k = \sigma(n)/n$. To prove this, divide both sides of $d_k + d_{k-1} + \cdots + d_1 = \sigma(n)$ by n.

17. If $x + y = a$ and $x - y = b$, then $ab = N$, and a can have $2d(N)$ different values: the positive divisors of N and their negatives. Since N is odd, so are a and b, and hence each pair (a, b) gives a distinct solution (x, y).

19. $\sigma_k(p^e) = 1 + p^k + p^{2k} + \cdots + p^{ek}$ and $\sigma_k(p_1^{e_1} p_2^{e_2} \cdots p_r^{e_r}) = \sigma_k(p_1^{e_1}) \sigma_k(p_2^{e_2}) \cdots \sigma_k(p_r^{e_r})$.

Section 9

9. The sum of the positive integers less than n and relatively prime to it is $n\,\phi(n)/2$.

15. n is of the form 2^a with $a \le 3$, $2^a p$ with $a \le 2$, or $2^a pq$ with $a \le 1$, where p and q are odd. Any other form would have $\phi(n)$ divisible by an odd prime or by 8.

17. Put $m = 2^r M$ and $n = 2^s N$, where M and N are odd and one of r and s is 1. Then $(M, N) = 1$, so

$$\phi(mn) = 2^{r+s-1}\phi(M)\phi(N)$$

while

$$\phi(m)\phi(n) = 2^{r-1}\phi(M)\,2^{s-1}\phi(N).$$

19. n is divisible by 6, so $n = 2^a 3^b N$ with $(2, N) = (3, N) = 1$. Thus

$$\phi(n) = 2^a 3^{b-1}\phi(N) \le 2^a 3^{b-1}N \le n/3.$$

Section 10

11. From Problem 10, $g = h^k \pmod{p}$ with k odd, so

$$(gh)^{(p-1)/2} \equiv (h^{(k+1)/2})^{(p-1)} \equiv 1 \pmod{p},$$

and gh does not have order $p - 1$.

13. Since $a \not\equiv 1 \pmod{p}$, $a^2 + a + 1 \equiv 0 \pmod{p}$, whence

$$(a + 1)^3 \equiv a^3 + 3a^2 + 3a + 1 \equiv 3(a^2 + a + 1) - 1$$
$$\equiv -1 \pmod{p}.$$

Thus the order of $a + 1$ is a divisor of 6. It is easy to verify that it is not 1 or 2.

15. We know that $a^2 \equiv -1$, so $a^3 \equiv -a$ and $a^4 \equiv 1 \pmod p$. Thus

$$(a + 1)^4 \equiv a^4 + 4a^3 + 6a^2 + 4a + 1 \equiv 1 - 4a - 6 + 4a + 1$$
$$\equiv -4 \pmod p.$$

17. From Problem 14, the primes other than 3 that can divide $2^{19} + 1$ are of the form $38k + 1$. Since $((2^{19} + 1)/3)^{1/2} < 419$, the only primes to test are 191 and 229.

19. Let $x = g^k$. Then

$$gx \equiv x + 1 \pmod p \quad \text{and} \quad g^2 x \equiv x + 2 \pmod p,$$

or

$$(g - 1)x \equiv 1 \pmod p \quad \text{and} \quad (g^2 - 1)x \equiv 2 \pmod p.$$

The latter congruence implies

$$(g + 1)(g - 1)x \equiv (g + 1) \cdot 1 \equiv 2 \pmod p,$$

which is impossible.

Section 11

15. $7 | (n^2 + 1)$ implies $n^2 \equiv -1 \pmod 7$, but this is impossible because $(-1/7) = -1$.

Section 12

1. If $p > 3$, then those integers k such that the least residue $\pmod p$ of $3k$ is greater than $(p - 1)/2$ are just those k such that $(p - 1)/6 < k \le (p - 1)/3$. Consider cases: $p = 12n + 1, 5, 7,$ or 11.

3. Show that if $p = 2^q - 1$, where q is an odd prime, then $p = 2 \cdot 4^{(q-1)/2} - 1 \equiv -1 \pmod 4$ so $(3/p) = -(p/3)$.

5. (a) Note that $q \equiv 1 \pmod 4$ always.
 (b) Consider two cases: $p \equiv 1$ and $p \equiv 3 \pmod 4$.

9. (a) $(-3/p) = (-1/p)(3/p)$ and apply Problem 1.
 (b) If $x^2 + xr + r^2 \equiv 0 \pmod p$, then $(x + s)^2 \equiv -3s^2 \pmod p$, where s is the unique solution of $2s \equiv r \pmod p$.

Section 13

15. Write 100000 in the base 2 (11000011010100000), and it is possible to get 100000 as a sum of powers of two.

19. (c) Suppose that an integer has two representations. Then

$$0 = (d_1 - e_1) + (d_2 - e_2) \cdot 2! + (d_3 - e_3) \cdot 3! + \ldots$$

Consider the equation (mod 2!) to see that $d_1 = e_1$, then (mod 3!) to see that $d_2 = e_2$, and so on.

Section 14

ϵ. The digits in the following answer are in base χ. One do-metric mile is 1728/1760 ordinary miles, one do-metric pint is 1616/1728 ordinary pints, and one do-metric pound is 27(62.5)/1728 ordinary pounds, using 62.5 as the number of pounds of water in a cubic foot and 231 cubic inches per gallon.

Section 15

13. (d) In base b, $1/(b - 1)^2 = .\overline{012 \ldots (b - 4)(b - 3)(b - 1)}$.

15. Yes. No.

Section 16

15. If $(a - 1)^2 + a^2 = (a + 1)^2$, then $a(a - 4) = 0$.

17. (a) $(2n + 1)^2 + (2n(n + 1))^2 = (2n(n + 1) + 1)^2$, $n = 0, 1, \ldots$.

(b) If $a^2 + b^2 = (b + 1)^2$, then $a^2 = 2b + 1$. Thus a is odd, say $a = 2n + 1$. Then $b = 2n(n + 1)$.

Section 17

3. If $x^8 + y^8 = z^8$, then $(x^2)^4 + (y^2)^4 = (z^2)^4$, and that is impossible.

Section 20

7. (c) The equation becomes $x + ay = 1$ or -1, and either one has infinitely many solutions.

9. (a) The area of a triangle with sides a, b, and c is

$$(s(s - a)(s - b)(s - c))^{1/2},$$

where $s = (a + b + c)/2$, so this triangle has area $(3a^2(a^2 - 1))^{1/2}$.

(b) If $3(a^2 - 1) = b^2$, then $3 b$, $b = 3c$, and $a^2 - 3c^2 = 1$, with solutions (2, 1), (7, 4), (26, 15), \ldots.

Section 21

1. (b) $[n/2] + [n/2^2] + \cdots < n/2 + n/2^2 + \cdots = n$.

3. p is not a factor in the denominator and is a factor in the numerator, once.

5. $n = \pi(p_n) \leq a p_n/\ln p_n \leq a p_n/\ln n$, because $p_n \geq n$.

Section 23: Additional Problems

Section 2

5. Suppose that n is composite. Then it has a prime divisor among p_1, p_2, . . . , p_k. Call it p. Then $p \mid (a - b)$ and $p \mid (a + b)$, so $p \mid 2a$ and $p \mid 2b$, which is impossible because $(a, b) = 1$.

Section 4

3. (b) The number is congruent to 2 (mod 9).

Section 5

3. For those k such that $(k, k(k + 1)/2) = 1$, and this is not true if $k \geq 3$.

Section 6

1. (a)
$$(ab)(ab)' \equiv 1 \equiv 1 \cdot 1 \equiv (aa')(bb') \equiv (ab)a'b' \pmod{p};$$
 cancel ab.

3. This can be found from the theorem of Problem 14, Section 6.

7. $(2p - 1)! \equiv p \pmod{p^2}$.

9. $4 \equiv -p(p + 1)p \pmod{p + 2}$, so
$$4((p - 1)! + 1) \equiv -p(p + 1)p((p - 1)! + 1)$$
$$\equiv -p((p + 1)p(p - 1)! + (p + 1)p)$$
$$\equiv -p((p + 1)! + 2) \pmod{p + 2}.$$

Thus $4((p - 1)! + 1) + p \equiv -p((p + 1)! + 1) \pmod{p + 2}$, and this is zero if and only if $(p + 1)! + 1 \equiv 0 \pmod{p + 2}$. That is so if and only if $p + 2$ is prime.

Section 8

5. An integer between a pair of twin primes greater than 5, 7 is a multiple of 6, and 12, 18, 24, 30, . . . are all abundant.

13. The divisors of $1 + p$ are included among $1, 2, \ldots, p - 1, p + 1$, and the sum of these is $(p^2 + p + 2)/2 < p^2 + p + 1$.

Section 9

5. This is essentially the same as Problem 19 of Section 9.

9. Let $n = 2^k 3^j N$ with $(2, N) = (3, N) = 1$ and consider cases. If both k and j are positive,
$$2^k 3^{j-1} N = n/3 = \phi(n) = 2^k 3^{j-1} \phi(N),$$
whence $N = 1$. If k is zero and j is not,
$$n/3 = 3^{j-1} N = 2 \, 3^{j-1} \phi(N),$$
which is impossible since $2 \nmid N$. If $j = 0$, then $n/3$ is not an integer.

13.
$$\sum_{d\mid n} (-1)^{n/d}\phi(d) = \begin{cases} -n \text{ if } n \text{ is odd} \\ 0 \text{ if } n \text{ is even.} \end{cases}$$

Section 10

7. If $m \geq n$, the result follows by multiplying both sides of $a^m \equiv a^{m-n}a^n$ (mod p) by a^{-n}. If $m < n$, start with $a^n \equiv a^{n-m}a^m$ (mod p).

11. Let $\text{ind}_g\, a = r$. Then $g^r \equiv a$ (mod p), so $a^n \equiv g^{rn}$ (mod p). But this says that $\text{ind}_g\, a^n \equiv nr$ (mod $p-1$), which was what was to be shown.

15. The first part is Problem 20(a) of Section 10 again. For the second part, see Roberts [14, p. 294].

Section 11

9. If $p \equiv 7$ (mod 8), then $q = (p-1)/2 \equiv 3$ (mod 4), so
$$(q/p) = -(p/q) = -(1/q) = -1.$$

Section 12

1. One of p, $4p + 1$, and $16p + 5$, is divisible by 3 for all p.

3. (a) The proposed equation is impossible (mod 3).
 (b)
$$\begin{aligned} m^2 &= n^2 + (n+1)^2 + \cdots + (n+k)^2 \\ &= (k+1)n^2 + k(k+1)n + (1^2 + 2^2 + \cdots + k^2) \\ &\equiv 1^2 + 2^2 + \cdots + k^2 \text{ (mod } k+1). \end{aligned}$$

5. $n^{2^m} \equiv n^{p-1} \equiv 1$ (mod p), and
$$-1 = (n/p) \equiv n^{(p-1)/2} \equiv n^{2^{m-1}} \text{ (mod } p),$$
so n has order 2^m (mod p).

Section 13

1. (c) $a(1 + b + b^2) = b(1 + c)$ implies $b \mid a$. But $0 < a < b$.

5. $1 = -1 + 1 \cdot 2 = -1 + (-1) \cdot 2 + 1 \cdot 4 = \ldots$.

Section 14

5. It is like the test in Additional Problem 10, Section 4.

Section 16

1. If $n + (n+1) = m^2$, then $n^2 + m^2 = (n+1)^2$.

3. If $n = t(t-1)/2$ and $m = t(t+1)/2$, then $m^2 - n^2 = t^3$.

5. If $mn(m^2 - n^2) = k(2mn + (m^2 - n^2) + (m^2 + n^2))$, then $2k = n(m - n)$. If $2k + ab$, then $m = a + b$ and $n = a$.

Section 17

5. If $p \nmid xyz$, then $x^{p-1} \equiv y^{p-1} \equiv z^{p-1} \equiv 1$ (mod p).

7. Yes, because $(n^2, n+1) = 1$.

Section 18

3. Suppose that $4^e(8k + 7) = x^2 + y^2 + z^2$. Apply Problem 1 e times to get $8k + 7 = x_1^2 + y_1^2 + z_1^2$. Then apply Problem 2.

5. If $n \equiv 2$ (mod 4), then $n = x^2 - y^2$ is impossible.

7. If $n = x(x + 1)/2 + y(y + 1)/2$, then $4n + 1 = (x + y + 1)^2 + (x - y)^2$.

9. None—all are congruent to 3 (mod 4).

Section 19

1. $5725841 = 11^2 \cdot 47321$, so it is necessary only to write 47321 as a sum of four squares.

5. If k_1, k_2, \ldots, k_r are odd, then $k_1^2 + k_1^2 + \cdots + k_r^2 \equiv r$ (mod 8).

Section 20

1. The approximations are 3/2, 17/12, 99/70, and 577/408.

3. This follows from Lemma 2.

Miscellaneous Problems

9. The first numbers on the left-hand sides are every other triangular number. The result may be written

$$(2n^2 + n)^2 + \cdots + (2n^2 + 2n)^2 = (2n^2 + 2n + 1)^2 + \cdots + (2n^2 + 3n)^2.$$

13. $f(n) = (3 + (-1)^{n+1})/4$ is one. $f(n) = (n + 1 - 2 [n/2])/2$ is another.

19.

	Now	Then
Ann's age	a	$3m/2$
Mary's age	m	$5a$

$3m/2 - a = 5a - m$, so $5m = 12a$.

27. $x^2 + (x + 1)^2 = (x(x + 1) + 1)^2 - (x(x + 1))^2$.

29. No: $x^2 + (1 - x) = x + (1 - x^2)$ for all x.

31. (b) $2a \equiv a^3 \equiv 3$, $3a \equiv a^4 \equiv 4$ (mod p) imply $a \equiv 1$ (mod p), which implies $p = 1$.

33. (a) If m is composite, then one of $p = 2, 3, 5, 7$ divides m. But then $p \,|\,(210n + m)$ too.

35. Note that there is no loss of generality in assuming that $(a, b) = 1$. Complete the square on the right-hand side to get

$$(x + (a + b))^2 = 2(a^2 + b^2 + ab),$$

and that is impossible (mod 4).

39. All are congruent to 1 (mod 3). If not, two of the integers would be divisible by 3.

41. $c^2 + 5 = cd + 3d$, so $d = c - 3 - 14/(c + 3)$, whence $c = 4$ or 11.

43. Any divisor is $2^r q^s$ where $r \le p - 1$, $s = 0$ or 1, and $q = 2^p - 1$.

47. Interchange 49 and 94.

49. There are $3 \cdot 365 + 366 = 1461$ days between one leap year February 1 and the next. $1461 \equiv 5 \pmod 7$. Remember that 2000 will be a leap year.

51. (a) If n is composite, $10^n - 1$ is composite.
 (b) $111 = 3 \cdot 37$.

53. Put $x = N + a$ and $y = N + b$. Then $ab = N^2$. There are $d(N^2)$ positive values of a that satisfy this and $d(N^2)$ negative values, one of which is $-N$.

55. (c) Let $\psi(n)$ be the number of elements in the sequence $1 \cdot 2, 2 \cdot 3, \ldots$, $n(n + 1)$ which are relatively prime to n. If $n = p^k$, p an odd prime, then $\psi(n) = p^{k-1}(p - 2)$. ψ is multiplicative; thus $\psi(n)$ can be found for any positive integer n.

59. For example, $19 = 16 + 2 + 1$ and 19 appears in lists 16, 2, and 1.

61. The harmonic mean of the divisors of n is

$$\left(\frac{1}{d(n)} \sum_{d \mid n} \frac{1}{d} \right)^{-1},$$

which is $nd(n)/\sigma(n)$. If $n = 2^{p-1}(2^p - 1)$, $\sigma(n) = 2n$, $d(n) = 2p$, and the harmonic mean is p.

63. $2r^2 + 3 \equiv 3$ or $5 \pmod 8$, which is impossible.

69. $9^{10} \equiv 1 \pmod{100}$.

71. $3 \mid (2^m + 1)$.

73. Yes: $2^{2k-2}(2^{2k-1} - 1) = 1^3 + 3^3 + \cdots + (2^k - 1)^3$ is true for all k, as may be shown using the fact that

$$1^3 + 2^3 + \cdots + n^3 = (n(n + 1)/2)^2.$$

75. If $n = 2^{p-1}q$, where $q = 2^p - 1$ is prime, then the divisors of n are

$$1, 2, \ldots, 2^{p-1}, q, 2q, \ldots, 2^{p-1}q,$$

and their product is n^p.

77. For each 5 that appears as a factor of $n!$, there are at least two even factors.

79. (a)

n	2	3	4	5	6	7	8	9	10	11	12	13	14	15
$f(n)$	2	3	4	5	3	7	6	6	5	11	4	13	7	5

n	16	17	18	19	20
$f(n)$	8	17	6	19	5.

83. If $p > 12$, then $p - 9$ is even and composite.

85. If $9 = (a/c)^2 + (b/c)^2$, then $a^2 + b^2 = 9c^2$. Hence $a^2 + b^2 \equiv 0 \pmod 9$, and

this implies that $a = 3r$, $b = 3s$. Thus $r^2 + s^2 = c^2$, and this has infinitely many solutions.

91. For $n \geq 2$, $x = y$ is impossible, so we may assume that $x > y$. Then

$$(x + 1)^n > x^n + nx^{n-1} > x^n + y^n > z^n > x^n,$$

so $x + 1 > z > x$, which is impossible.

93. If $f(r/s) = 0$, then $r \mid a_n$ and $s \mid a_0$. Hence r and s are odd. But this implies that $0 = s^n f(r/s)$ is the sum of some even integers and an odd number of odd integers, which is impossible.

95. Induction, using the identity

$$2^{3^{k+1}} + 1 = (2^{3^k} + 1)(2^{2 \cdot 3^k} - 2^{3^k} + 1),$$

is one method that will work.

97. (a) Let $m = p_1 p_2 \ldots p_k$ with $p_1 < p_2 < \cdots < p_k$. Then $p_2 \mid m$ implies $(p_2 - 1) \mid m$, so $p_2 - 1 = p_1$. Thus $p_1 = 2$ and $p_2 = 3$. Further, $p_3 \mid m$ implies $(p_3 - 1) \mid m$, so $(p_3 - 1) \mid p_1 p_2$, whence $p_3 = 7$. Similarly, $(p_4 - 1) \mid 42$, so $p_4 = 43$. Finally, $(p_5 - 1) \mid 2 \cdot 3 \cdot 7 \cdot 43$, but there is no such prime.
 (b) $m = 2 \cdot 3 \cdot 7 \cdot 43 = 1806$ has the desired property.
 (c) But there are no others.

99. Apply Fermat's Theorem: there is a progression starting at any term with ratio 2^{p-1}.

Appendix A

3. $1^3 + 3^3 + \cdots + (2k - 1)^3 = k^2(2k^2 - 1)$.

5. $t_n = n(n + 1)/2$.

9. $(n - 1)n(n + 1)(n + 2) = (n^2 + n - 1)^2 - 1$.

13. $8t_n + 1 = (2n + 1)^2$.

15. $f_{5n+5} = f_{5n} + 5f_{5n-1} + 10f_{5n-2} + 10f_{5n-3} + 5f_{5n-4} + f_{5n-5}$.

Index